Volume 5

SOCIAL RESEARCH TECHNIQUES
FOR PLANNERS

SOCIAL RESEARCH TECHNIQUES FOR PLANNERS

THOMAS L. BURTON AND
GORDON E. CHERRY

Routledge
Taylor & Francis Group

LONDON AND NEW YORK

First published in 1970 by George Allen and Unwin Ltd

This edition first published in 2018
by Routledge
2 Park Square, Milton Park, Abingdon, Oxon OX14 4RN

and by Routledge
711 Third Avenue, New York, NY 10017

Routledge is an imprint of the Taylor & Francis Group, an informa business

© 1970 Thomas L. Burton and Gordon E. Cherry

British Library Cataloguing in Publication Data
A catalogue record for this book is available from the British Library

ISBN: 978-1-138-49611-8 (Set)
ISBN: 978-1-351-02214-9 (Set) (ebk)
ISBN: 978-1-138-48770-3 (Volume 5) (hbk)
ISBN: 978-1-138-48781-9 (Volume 5) (pbk)
ISBN: 978-1-351-04222-2 (Volume 5) (ebk)

Publisher's Note
The publisher has gone to great lengths to ensure the quality of this reprint but points out that some imperfections in the original copies may be apparent.

Disclaimer
The publisher has made every effort to trace copyright holders and would welcome correspondence from those they have been unable to trace.

Social Research Techniques for Planners

Thomas L. Burton

Associate Professor of Urban and Regional Planning, University of Waterloo, Ontario; formerly Centre for Urban and Regional Studies, University of Birmingham

and

Gordon E. Cherry

Deputy Director, Centre for Urban and Regional Studies, University of Birmingham

London
GEORGE ALLEN AND UNWIN LTD
RUSKIN HOUSE · MUSEUM STREET

FIRST PUBLISHED IN 1970

© Thomas L. Burton and Gordon E. Cherry, 1970
ISBN 0 04 711002 3

PRINTED IN GREAT BRITAIN
in 10 on 11pt. Times New Roman type
BY UNWIN BROTHERS LIMITED
WOKING AND LONDON

Acknowledgements

This volume may be considered as a direct product of our work at the Centre for Urban and Regional Studies, University of Birmingham. Although our previous experience in other capacities has also materially contributed to the scope and content of the work, it is true to say that without the emphasis on social science research teaching and application at the Centre, and our own involvement in it, we should not have been persuaded to undertake this venture.

We have therefore to give sincere thanks to the Director, Professor J. B. Cullingworth, not only for his helpful advice during the drafting stage of this volume, but also for the very considerable weight he attaches to the teaching of social science research, an emphasis to which we also would strongly subscribe. We must additionally thank our many colleagues at the Centre who, through hours of self-examination, have considered and reconsidered the form of our postgraduate teaching, and have always been the source of most stimulating views on particular courses. In this respect we must single out Miss Cilla Noad who contributed directly to the material in several chapters. From our first batch of students in the session 1968-9 there was a valued feed-back of constructive criticism, and our recognition of this should be recorded.

There is, of course, one group of people to whom we owe the greatest debt of all: the many research scientists and writers whose work and studies have provided the raw material for this book. It is an unfortunate fact that when a subject has advanced to a stage where there is a sufficient body of basic knowledge to constitute a textbook, the work of the many pioneers, the researchers who, little by little, have created that body of knowledge, tends to be forgotten in favour of those who put it all together. Perhaps it is only those who put it together who can ever fully realize the debt that is owed to those working on the frontiers of the subject. To them, we offer our grateful thanks and our profound respect. A list of the major works from which we have drawn is given in the Bibliography.

Finally, we should like to thank our respective secretaries, Miss Mary Grant and Mrs Jacky Bulgin, for their patience, hard work and skill in presenting a manuscript to our publishers with the speed which was demanded.

Preface

The purpose of this volume is to provide an introduction to social research methods for the town planner. The first chapter presents the case for a greater emphasis on social research within the planning process. The remaining chapters, concerned with the kind of problem faced by the research worker at successive stages of a research project, outline the major conceptual and organizational problems likely to be encountered in any social research for planning, with suggested guidelines for tackling these.

The underlying theme of the book is very relevant to the full-time graduate course for the Diploma in Urban and Regional Studies recently instituted at the Centre for Urban and Regional Studies in the University of Birmingham, and springs from a series of lectures prepared for this. The teaching for the Diploma is focused upon social science research methods and the application of social and economic research to issues of urban and regional planning and administration. The purpose is to provide a training for future research workers, not just for professional town planners but for administrators and other members of planning and related professional teams, both within and outside government.

The teaching of research has been neglected for too long, and for the town planner, as with others, this is now serious. It is no longer sufficient to consider research method as something that can be 'picked up' in the course of actual research projects. Too many large and alarming mistakes are made in this way, and too often valuable data, often acquired at considerable cost in time and effort, prove to be unusable because, owing to ignorance of basic methodology, they have been collected unsatisfactorily. In these circumstances we believe that the Diploma in Urban and Regional Studies will be a valuable qualification for the research specialist in a planning team.

But the numbers of people who can follow this course are, for practical reasons, severely limited. Planners without this particular training still need to be familiar with basic research techniques, and it is in order to help to meet this requirement and to stimulate the interest of both the specialist and the non-specialist, planner, professional and non-professional, that this volume has been conceived.

THOMAS L. BURTON
GORDON E. CHERRY

Centre for Urban and Regional Studies
University of Birmingham
November, 1969

9

Contents

Tables

Chapter 1

SOCIAL RESEARCH AND THE TOWN PLANNER

Town planning in twentieth century Britain has been concerned from the outset not only with physical but also with broadly social aspects of town life.

On the one hand, there was the drive for sanitary reform which provided the basis for substantial urban improvement and in the quality of town and city life. At the end of the century the rediscovery of the artistic principles of town building continued the world-wide tradition of planning from the point of view of architecture and urban design. On the other hand, the Victorian social reformers, represented in mid-century by theorists such as Buckingham with his proposal for a Model Town, Victoria (*National Evils and Practical Remedies*, 1849) and practical exponents such as Salt at Saltaire with model housing in conjunction with a relocated manufacturing concern, were important figures in the drive for housing improvement. Towards the end of the century, there were examples of superior working class housing at Bournville by Cadbury, at Port Sunlight by Lever, and later at Earswick (Yorks.) by Rowntree. These developments and the publication in 1898 of Ebenezer Howard's *Tomorrow: a Peaceful Path to Real Reform* (revised as *Garden Cities of Tomorrow*, 1902) led directly to the Garden City movement and the first experiment at Letchworth in 1903.

This social background to planning became manifest in two main directions. First there were the related questions of house design and the provision of improved facilities; the arrangement of dwellings which afforded the prized amenities of light, air and space; and the layout of residential areas which give the opportunity for a locally based community life. Secondly, under the influence of an anti-urban viewpoint which considered city life in the light of excessive overcrowding and the related social evils which had been exposed, there was the desire to control urban form; this was particularly so with regard to big cities, especially London, where rapid suburban

15

expansion had been a feature of the late nineteenth century. It was this latter aspect that gave the planning movement its first legalistic framework (the Housing, Town Planning etc. Act 1909): permissive powers over the control of land use were tentatively extended to the selection of housing land and the manner of its development.

The social factors which affected the development of British town planning were part of a wider background, and in particular it is relevant to note the increasing attention devoted to social enquiry at the end of the nineteenth century. Henry Mayhew's *London Labour and the London Poor*, 1861, was a forerunner, and increasing attention was devoted to Census returns. But towards the end of the century a number of significant reports supported the drive towards better housing. At an emotional level there was the vivid tract by a Congregational minister, Andrew Mearns, *The Bitter Cry of Outcast London*, 1883, and General William Booth's survey with proposals for city, farm and overseas colonies, *In Darkest England and the Way Out*, 1890. More soberly there was Seebohm Rowntree's *Report of the Royal Commission on the Housing of the Working Classes*, 1884–5, and subsequently his survey at York.

But the most notable example was the empirical research of Charles Booth. Volume 1 of *Labour and Life of the People* (East London) was published in 1889, and Volume 2 (London Continued) appeared in 1891. These findings of his study of the relationship between employment and poverty among London's working classes contributed materially to the anti-urban tradition of the day, to be expressed in the view that London was too big and that the only solution lay in developing satellite communities away from the urban mass. Subsequently nine volumes of *Life and Labour of the People in London* appeared between 1892 and 1897, covering industry and religion.

This spirit of enquiry gave rise to a close relationship between the developing social sciences and town planning. Early work provided evidence of poor housing and its consequences, and the association of sub-standard accommodation and high densities with the occurrence of crime and ill-health gave a tremendous stimulus to the preparation and execution of the first planning schemes. The early town planners were committed social reformers, firmly believing, in the spirit of environmental determinism, that their new housing areas would help to eradicate the social evils of the day. This social base of town planning has changed substantially but has not been eroded; the evolution of planning in both theory and practice has continued to owe much to social study. For example, one of the concerns of the

16

early twentieth century sociology was the organization of community; town planners found an expression in the Garden City principle which offered much promise. Raymond Unwin expressed it thus: 'This basic principle of organization should find its expression in the form of the town which, instead of being a huge aggregation of units ever spreading further and further away from the original centre and losing all touch with that centre, should consist of a federation of groups constantly clustering around new subsidiary centres, each group limited to a size that can effectively keep in touch with and be controlled from the subsidiary centre. . . .'[64]

Patrick Geddes saw the potential of this relationship between planning and the social sciences: 'the renewed art of Town Planning has to develop into an art yet higher, that of City Design – a veritable orchestration of all the arts, and correspondingly needing, even for its preliminary surveys, all the social sciences'.[21] More specifically, 'we need to search into the life of city and citizen, and the interrelation of these, and this as intensively as the biologist enquires into the interaction of individual and race evolution. Only thus can we adequately handle the problems of social pathology; and hence again rise to the hope of cities and with clearer beginnings of civic therapeutics, of social hygiene.'

It is not appropriate in this volume to give a detailed indication of the contribution of social enquiry to town planning in the twentieth century. But in order to establish the point that this relationship has been quite fundamental in recent planning history, and is likely to continue to be so, we will briefly touch on the more important developments that have occurred.

Firstly there was the substantial progress in the understanding of the urban environment which Geddes had urged: this was R. E. Park's contribution to urban sociology during the 1920s and 1930s in Chicago. In particular, the attention devoted to the processes of invasion and succession, the monographic studies of city life such as *The Gold Coast and the Slum* and *The Ghetto* and the emphasis on ecological analysis in the study of urban spatial patterns assisted materially in theoretical thinking about urban structure. This was to lead to further studies which have provided the planner with a theoretical basis for the understanding of towns and cities. First of all Burgess' concentric zone theory was developed in the early 1920s to explain ecological processes.[5] In the next decade this led to Hoyt's study of residential areas in the United States which gave a new theoretical explanation of residential land use, accounting for wedge-shaped sectors radial to a city centre.[32] In post-war years enquiry and hypothesis in this field have continued to provide new interpretations and these are aids in the evolution of planning practice: Firey's work in Boston pointing to the importance of cultural ecology[17] and that of Grigsby in Philadelphia concerning the interrelationship of sub-markets in the total housing system have led to the concept of 'filter' in residential areas, a process whereby in twilight areas ageing and unfit housing is passed down the social scale. Town planning needs theory on which to base, and against which to test, its practice; the results of ecological analysis have made an outstanding contribution.

Another example of social enquiry assisting in the development of town planning thought was concerned with the question of 'community'. Here, the neighbourhood theory made a contribution in practice to the rational ordering of urban spatial patterns. Clarence Perry's important paper *The Neighbourhood Unit, a Scheme of Arrangement for the Family-Life Community* was published as part of the New York Regional Survey of 1929, but contributory ideas were already of long standing. For example Cooley had stressed the geographical basis of primary formative association.[14] Moreover, the community centre movement was firmly established. The settlement house movement originating at Toynbee Hall, London, in 1885 was adopted in the United States, notably by Jane Addams on Chicago's Halstead Street. The movement did not however confine itself to blighted areas, but aimed, rather, to organize any neighbourhood's public and private activity in the fields of recreation, general culture and adult education. This tradition was to lead to the emphasis on neighbourhood planning which so characterized British policy from the *Greater London Plan*, 1943, to the standard Development Plans of local authorities throughout the 1950s.

Next, we might refer to the contribution to contemporary planning by a large number of community studies both in this country and the United States. These have facilitated a deeper understanding of particular communities as well as interrelationships between sets of institutions in particular localities. In this tradition we might note *Coal is our Life*, the study of a Yorkshire mining community with a focus on workplace, *Small Town Politics*, a study of Glossop concerned with political institutions, *The Sociology of an English Village* which dealt with the question of land and an interrelationship with family and class, and *The Family and Social Change*, studying the family in Swansea. A most influencial publication was Willmott and Young's *Family and Kinship in East London*. This described working class life in Bethnal Green comparing it with that in an overspill situation, 'Greenleigh'. There has been a large number of subsequent studies, and one of the impacts of these has been that they have contributed substantially to a new consideration of slum clearance and the disruption of local, long-established communities. Planning and related housing practice is now better informed as a result of studies in this field, and there is now a much greater awareness that intervention by a local authority in the housing market can be a very insensitive instrument of social policy.

Another aspect of community studies to have a bearing on planning is that which examines the relationship between the physical and social environments. That by Cherry in Newcastle is a recent example with regard to the distribution of 'social malaise'.[9] A relationship between poverty of environment, and especially overcrowding, and the incidence of a range of physical, mental and social 'ill health' can readily be demonstrated, but any conclusion as to a *causal* association must be very tentative. The key characteristic is the interrelationship of all the factors affecting 'behaviour', making the identification of any particular influence extremely difficult. This is important, and its relevance is to be seen in the final erosion of the planner's former reliance on environmental determinism. Planning inherited from the Victorian reformers the idea that the physical environment was a major determinant of society and culture, and consequently that if the physical environment were improved man would gain substantially in happiness.

As a concept, physical determinism is now being replaced by a much broader 'systems' approach. As Gans writes, social scientists point to a more complex relationship than that previously held. 'Aided by research findings which indicate that the portions of the physical environment with which city planners have traditionally dealt do not have a significant impact on people's behaviour and by studies of social organization and social change which demonstrate that economic and social structures are much more important than spatial ones, the rational programmers devote their attention to institutions and institutional change rather than to environmental change.'[20] This new view of the planning process confirms the importance of the relationship with the social sciences.

It will be seen therefore that the evolution of town planning theory and practice owes much to the spirit of social enquiry and the development of the social sciences. This relationship shows no sign of termination; indeed, it would be strange if the close association were to be severed, for this would seem to deny the social basis of the planning movement. It is a relationship, however, which planners may find uncomfortable because of unsureness in handling the methods of social survey; hence the need for this volume.

19

Problems for the Planner: Education and Interdisciplinary Study

At the present time new attention is being paid to the social aspects of planning: there is a reformulation of views both as to social and planning goals and the social consequences of the (largely 'physical') planning process. In view of the socially-oriented nature of the origins of the planning movement such a rethinking is to be welcomed. Indeed, it might be considered unduly belated, but this is in part due to the late fashioning of the 'tools of the trade' of contemporary social enquiry. The development of new techniques and aids has allowed social research in this country to expand considerably in the last two decades.

Planners have tended to rely in the past on the physical sciences rather than the social sciences and they are still slow to recognize the importance and relevance of social surveys. This situation is not helped by the fact that planning practice in government offices is problem-oriented: short-term research activity, where work programmes take on the character of 'trouble-shooting' with surveys designed to clarify particular problems, has not attracted the financial or manpower resources necessary for long-term social research. A second difficulty concerns planning training and the poor understanding of social research which has led to unfamiliarity about techniques and methodology. As a consequence, the practising planner has tended to neglect his thinking about the need for social survey as a fundamental part of his work; the truth of the matter is that he can no longer afford to do so.

Another problem for the professional planner is the way in which social research is difficult to isolate as a function or practice. For example, it is undertaken by a number of different agencies: local planning authorities, central government, university departments, consultants, market research organizations and supported by research trusts and foundations. Moreover it is carried out by people with varying academic and professional backgrounds, bringing to bear a wide range of technical skills and disciplines: town planners, sociologists, social administrators and social scientists of all kinds. Here indeed is a marked example of how research interests are being refocused on a range of studies, which increasingly fall in the previously uncharted borderland between traditionally accepted fields of interest.

Population migration, for example, might be studied by the demographer as a historical investigation or as an exercise in forecasting. Alternatively, the geographer might be interested from the point of view of a case study in regional analysis. Again, the social scientist might be more interested in the motivational aspects of the migrants and the study of attitudes and aspirations which underlie reasons for movement. The social administrator, on the other hand, might see mobility from the point of view of the social needs of the migrants – those destined for new communities, for example, where the planned provision of social facilities is so much required. Finally, the town planner is vitally concerned simply because the movement of people, whether over short distances or long, is associated with changing characteristics of residential environments and the planning problems with which they are associated: rural change, urban peripheral spread, the changing nature of communities, housing and the location of industry, to name but a few.

In most cases the studies carried out by the various investigators will be relevant to the interests of others; and, moreover, the manner in which the study is undertaken may well require advice or contributions from specialisms other than that of the originator. From the point of view of the town planner this is especially important because it means that many aspects of social research will be carried out by non-planners; but the findings have to be interpreted and worked upon by planners, if an ultimate contribution to planning policy formulation is to be the end product. The town planner will never carry out either all or even most of the social research, the findings of which he will want to use: he will always be very much dependent on the activities of others. This poses problems of interdisciplinary communication, the best use of scarce resources and the coordination of programmes for the determination of priorities.

Social Research in the Planning Process

The case for greater recognition of social research in the planning process may be argued on the basis of three important features of present-day planning. The first of these is the changing role of planning itself, from a concept of static plan-making to an evolutionary process in which plans are adapted in the light of changing circumstances. The second is that, as a consequence, planning is as concerned with implementation as with scheme preparation. In the carrying out of planning policies it is even more necessary to show that the objectives are meeting immediate as well as long-term needs; of these, we shall see that social needs are as important as any. The third is that the planner is concerned not only with land use patterns in formulating schemes for urban shape and form: his raw material is not just a physical environment but the social systems indissolubly linked with that environment. These three aspects deserve some elaboration.

Firstly, for much of this century the usual image of planning has been its concern with plan-making. Sometimes, ideal postulates have been presented in three-dimensional terms when the architect-planner conceived his solution as an act of creative townscape. For larger areas, schemes have been presented for the future form of towns and cities, the basis of which has commonly been improved and 'convenient' distribution of various major land uses. In view of the inherited disorder of Victorian cities this rational improvement in the disposition of land activities has justifiably been regarded as a major planning goal, and only the more zealous attempts at undue segregation have brought the principle into question.

The essence behind this tradition has been that the planner outlined an ideal solution which was largely static in concept: the conventional policies developed to incorporate, for example, satellite communities, a rationalized distribution of land uses, neighbourhood planning, an improved road pattern, the provision of open space and facilities to prescribed standards, and the adoption of housing layouts which varied over time according to design fashions and techniques. Plans have been presented as models for the future, the success of which was largely unquestioned because they were seen to overcome or remedy the deficiencies of the past and present; as far as the future was concerned, this was unknown, and if it were to be different from the present it would be suitably moulded to the new order. The rash of plan-making during the middle 1940s and, indeed, the main intention of the Development Plan as outlined in the Town and Country Planning Act, 1947, belongs to this traditional view. But as Duhl reminds us, 'Planning, in its truest sense, is not a static state; planning is not a product, but a process – a veritable evolutionary continuum. It is not creating a grand or master plan that must be implemented no matter what happens.'[16]

In the preparation of these plans a serious shortcoming was the reliance on the physical sciences, as against the social sciences, for the purposes of analysis of contemporary situations and predictions for the future. The techniques of survey tended to be concerned with physical factors such as population trends and housing needs, communications and land use. As far as sociological considerations were concerned there was a reliance on value judgements, notable examples being the subjective concern for community, and neighbourliness; it was difficult to distinguish between survey and policy and there were few attempts (although Ruth Glass's work at Middlesbrough is an obvious exception) at objective assessments of social needs and aspirations as part of plan-making.

The rejection of static plan-making in favour of an adaptive process (adaptive in the sense of modification to changing circumstances) is profoundly significant. Reliance on a series of changing short-term objectives within the framework of broad, long-term goals permits and, indeed, enforces the planner to be sensitive to ongoing change and to leave options open in crucial decision taking for as long a period as possible. There will constantly be evidence of new situations and circumstances; demographic data and a variety of social factors relating to housing choice, preferences and changing aspirations and attitudes will all be important. Moreover, there will be the need for the constant monitoring of the consequences of planning decisions or policies. In other words, opportunities and the need for social research in planning are extended as planning departs from a concern for the consolidation of fixed goals to a process which is dynamic and adaptive over time. Social research is an integral part of this process, which allows policies to be modified and redefined as necessary in the light of changing conditions.

Secondly, the evolution of twentieth century planning was closely associated with the wider social reform movement, and planning has as its base an objective which is to meet certain social needs and broad goals. These were formerly thought to be secured in paternalist fashion through a control of land use and a sensitive manipulation of the built environment, especially with regard to housing. A related assumption was that, in the building of new neighbourhoods or new towns or in the improvement of existing areas, adherence to certain standards of provision for such as open space, shops and community buildings at prescribed rates adequate for each neighbourhood would meet basic needs, provide satisfaction and constitute the social basis for a happy community.

Such a simple approach to the provision of facilities can rarely be completely satisfactory because the age-group and particular needs of communities vary; moreover, situations change over time. Neighbourhoods have rarely developed the self-sufficiency that was expected of them by the more idealistic shapers of community life, and a considerable overlap in the catchment areas for various facilities and services between neighbourhoods is now recognized. These tributary areas for such as shopping, recreation and community services require constant reappraisal if provision of the necessary facilities is to be sufficiently realistic to give maximum consumer satisfaction.

Additional serious shortcomings are, of course, now recognized in that the social objectives of planning are met in a much wider manner than that resulting from land use and environmental manipulation. Questions of employment opportunity, coordination of educational, health, welfare and general community services are now becoming recognized as part of the planners' brief, and it is from improved planning in this direction that elements of basic community satisfaction might be provided.

The changing scope of social planning can also be seen in housing. Social satisfaction through the provision of better housing has been a long standing objective of planning. Such was the legacy of nineteenth-century development which gave rise to an unprecedented backlog of unfit housing in the middle of the twentieth century that the planner's interpretation of need has continued to be simply the provision of new and substantially improved accommodation. A similarity between the contemporary housing architect-planner and the nineteenth-century housing reformer is in the paternalist view: new facilities are provided according to the design whims of the day with remarkably little consumer research as to what forms of accommodation are required, simply motivated by a belief in the unquestioned desirability of improved or additional housing facilities and in the efficacy of their provision as an element in social improvement.

The shortcomings of this situation as far as social planning is concerned have been, firstly, that certain additions to the housing stock have been made which appear to be actually contrary to the spirit of meeting community needs: as, for example, in excessive high-rise development or in the lack of choice between high and low density or between types of accommodation, reflected particularly in the shortage of dwellings of larger size. More importantly, there is the recognition that housing improvement alone does not necessarily remove social problems. (It may in fact intensify them, if housing relocation results in high rent levels and family stress associated with financial difficulties.) As Gans observes, '... policies which seek to change the "physical" environment have little impact on the behaviour patterns and values of people. Planning which aims to improve living conditions must address itself to the significant causal elements of these conditions, which are usually economic, social and political.' [20]

The point being made here is that the mere provision of social facilities or improving housing does not by itself go very far in expressing the social objectives of planning. It is now being recognized that it is important to evaluate the relevant needs of communities as part of social policy making, and consequently this stresses the role of social research in planning. The revolution of thinking is that planning may be seen as an enabling process facilitating change and evolution in society through a host of individual creativities and actions, rather than as an ideal goal to which society is to conform. Therefore, the planner's concern is in meeting relevant needs and in fostering broad goals of society, quite as much as his accustomed role of shaping the physical environment.

The planner in this sense becomes the more complete servant of society, not merely the grand designer; planning which began as a reform movement is now a client-centred service. This is a tremendous extension of the traditional concept of the planner's role. For those who see this claim as excessive we can but recall that the planner has been concerned with environment: whereas this has usually been seen in physical terms, the meaning of environment extends beyond land and land use to towns and communities and people and social organizations in them. Planning is concerned with these communities and their needs as they are revealed. Not all social needs are central to the planner's interest of course; education and health, for example, are distinct fields of concern. But social research can help the planner to play a coordinating role for various social services and to act in an 'intelligence' capacity.

Thirdly, the reason for welcoming an extension of social research in planning is that the town which is the subject of the planner's concern is not just a physical environment but a matter also of interconnected social systems. As Pahl remarks, 'Cities are social entities and their physical characteristics only gain meaning when men give it to them'. [48] Physical urban changes, whether expansion or internal decay and renewal, stem from the sum of a myriad of individual desires and aspirations and a complex process of personal and group decision taking. For this reason, a planner should view an urban area as a social system in action quite as much as a physical artifact. This highlights the need for research into the various social aspects of cities, towns and rural areas to complement the more customary research into physical aspects of the environment. Knowledge of communities, the nature of community life, the relationship between the built environment and behaviour and response, all form relevant raw material for the planner.

An example of this can be illustrated with regard to the problems of twilight areas. Here the task is not to clear and redevelop, but to improve, revitalize and conserve dwellings and environment, and in so doing to allow for and meet the changing needs and aspirations of communities. It is increasingly clear that solutions to the problems of inner urban areas do not lie solely in physical improvement to prescribed standards. The adoption of uniformly similar techniques of housing and environmental improvement (whether or not by compulsory direction by the local authority) will certainly achieve something, but the process will set in train a chain of events throughout the whole housing market; the social aspects of this interaction need to be understood and monitored closely.

Many local authorities, under the stimulus first of the White Paper *Old Houses into New Homes* (Cmnd 3602, 1968) and subsequently of the Housing Act, 1969, seem ready to embark on programmes of physical renewal with the intention of providing rehabilitated environments for an extended life-span. It is true that the improvement of the stock of dwellings is very important, but this is a physical intervention which needs to be delicately handled in full knowledge of the social aspects of local situations. The arresting of urban decay is not just a process of eliminating or upgrading poor housing or obsolescent environments; it is concerned with the circumstances which force certain groups of people to live in areas of underprivilege.

By and large we know very little about twilight areas, but the usual hunch is that increasingly they are becoming 'residual' areas where poverty of living conditions is associated with a wider social deprivation. The young, energetic and upwardly mobile are thought to desert these areas for districts of living more in keeping with their assumed status and aspirations. The elderly, the low income groups and those permanently in a depressed social status tend to remain with others who are also disadvantaged in one form or another, but particularly from the point of view of housing – the immigrant, the large family, the poor, the newly-married and the single households whose dwelling choice is limited. It may be that the twilight ring has a function in that it provides low cost housing for precisely such as these people, and if as a result of planning intervention in the name of improvement, hardship ensues for substantial minorities because of misdirected paternalist idealism, then planning will be a blunt tool in shaping social change.

Certainly, a local authority has the obligation as part of social policy to improve its dwellings, and there are substantial economic arguments for seeking to prolong the life of residential properties and so obviate the necessity for early clearance. The difficulty lies in the delicacy and the understanding with which the improvement policy is carried out. This demands the closest attention to the wide social aspects of the whole operation; research is demanded so that the planner can shape his policies in fuller light of the underlying situations.

These then are the reasons for greater emphasis on social research within the planning process:

    1. The changing nature of plan-making: a move from a static

concept to one which is more an adaptive process where, in the need for greater understanding of situations, there is a reliance on social sciences as well as physical sciences;

II. The wider objectives of planning: if relevant social needs are to be met adequately by the planning process, an objective evaluation of requirements is necessary, with a more rigorous testing of inherited subjective criteria;

III. The concern of planners is not only with the physical nature of towns and land use. Because it is also with social systems and their interaction, the relevant research field is greatly extended.

As Gans conjectures, 'In its brief history, planning has developed from a missionary movement to a profession based on the beliefs of that movement, but with a strongly architectural and engineering emphasis . . . planning in its next phase would be a profession resembling in many ways the discipline of an applied social science. While it would depend on the social sciences for many of the data relating goals and programs, it would be an art in formulating its special synthesis of these data.'[20] This underlines the need for social research in planning and the desirability of planners having far greater knowledge of techniques and methods.

The Field of Social Research

The Town Planning Institute's *Register of Planning Research*[62] is a useful guide to the studies already undertaken by a number of local planning authorities and many other organizations. Furthermore, a broad review of the social studies which are of relevance to the planner's work is given by Cherry.[10] But it is helpful to summarize here in general outline the field of social research which provides so many contributions to planning.

We might suggest two main categories for the planner to bear in mind. First, there are 'general' aspects of research; these are studies which have a bearing on the broad understanding of social issues and trends, with no significant focus on any geographical base. Second, there are 'particular' aspects of research; these are studies which depend heavily on a given locality, urban, rural or regional, for the investigation of social issues, but which may be subsequently applicable in varying degrees to other areas of similar characteristics. The distinction between the two is sometimes difficult to draw, but it is one which it is still useful to make.

Under the first heading almost any project can be included which adds to an understanding of the social environment, and which is

27

relevant to the work of a planner. First there will be broad-based work which contributes to an understanding of contemporary social change. This will be concerned with, for example, the effect of urbanism or industrialism on society, the role of the family in modern society or movement between social classes (for instance, *embourgeoisement* of various skilled workers in the middle class) in a dynamic society where class barriers are in flux. Next, there are demographic studies; these would include the growth and movement of population over a given period of time, for the country as a whole or for broad divisions within it. Work on the relationship between environment and behaviour might also fall in the 'general' category, if we follow the analogous studies of scientists who have experimented, for example, with the effects of overcrowding on animals.

Then there are a number of fields where it is increasingly difficult to separate the 'general' from the 'particular', in that initial broad-based studies lead on as second or further stages to subsequent locally-based work. Of this type, research focused on industry might include studies of mobility of labour, journey to work, or the social features of specific industries and their dependent communities, such as coal mining or agriculture. Lastly, valuable 'general' background introductory work might be contributed to the question of underprivilege; for example, surveys relating to poverty, race relations, and problem families have been important in initiating later local studies.

The planner will frequently find that 'particular' studies are more helpful than 'general' ones, especially when the geographical locality in question is the subject of a planning exercise. In this category there is extensive material on which to draw, and for convenience a threefold review can be suggested: regional, urban and rural.

Regionally (or sub-regionally) based social research is rather limited, but the setting-up of regional units of administration has stimulated a demand for regional social data for planning purposes. Regional 'consciousness' or identity is perceived largely through social attributes such as dialect, cultural heritage, and 'way of life', and some of these can be highlighted, as for example by Allen.[2] But there is an increasing range of available socio-economic statistics, and their evaluation, as for example by Hammond,[29] can be very useful in drawing regional distinctions. Regional information on leisure and recreation is now available from two national studies[66] and regional additions to the stock of knowledge are being made.

At a sub-regional level a particular form of social investigation has been concerned with conurbation housing surveys by the Ministry of Housing and Local Government. These have provided an overall picture of housing conditions, housing needs and the potential and effective demand for housing.

Within urban areas the locally-based social survey and related case study has been of great importance for the planner. Analyses of the social characteristics of particular areas are imperative for the detailed planning of local communities; it will be necessary to obtain some data from original surveys, but otherwise Enumeration District Census material will commonly give an easily accessible introduction to area characteristics. Social area analysis is now recognized as an important planning tool because it allows, especially in cities, meaningful areas to be built up, without the constraint of administrative boundaries, on the basis of concentrations of particular socio-economic variables.

Urban area social studies can be of a wide variety. They can be studies of suburban communities, inner urban area communities, new towns or extensive settlements developed by one builder, each type varying because of geographical base, age, social characteristic or housing tenure; in fact, there can be as great a variety of these studies as the range of internal social and physical differences which the area in question can accommodate. Furthermore these studies can select particular issues rather than present the broad range. For example, there can be research into the evolving patterns of social relations on various kinds of estates, racial difficulties or the problems related to slum clearance and redevelopment.

Housing, in fact, will be a frequent area of study demanded by the planner. Work on the life of families in high flats or in particular types of designed layouts has contributed to a greater awareness of residential satisfaction with types of dwellings or environments. This has been supplemented by local residential mobility surveys which provide an understanding of social forces behind changing spatial patterns; the motivational aspects of migration may suggest underlying causes of satisfaction or dissatisfaction with particular types of residential district.

Housing is closely associated with a number of related social issues in urban areas. The question of environment and social response has supported studies of mental health on new estates and in old communities in the light of reported 'suburban neurosis' or 'new town blues'. On a wider basis of social health, identification of the distribution of 'social malaise' has been provided by the collection of various indices relating to crime, delinquency, educational subnormality, and problem families; these contribute to the recognition and identification of social priority areas.

The planner has been accustomed to providing social and community facilities on a neighbourhood basis to certain set standards. Few of the assumptions relating to standards of provision have been rigorously tested in the light of changing circumstances, and there is little reliable evidence so far on which to plan for a pattern of overlapping catchment areas for various facilities. Lee's work[37] on centres for further education and the optimum provision and siting of social clubs has pointed the way for further contributions.

The use of open space has probably been surveyed more than any other planned facility. The distribution of children's playgrounds has been suggested by a national survey[15] but more local surveys of the use of urban parks are required to help in detailed planning proposals. In most urban areas the National Playing Fields Association standard of six acres of playing fields per thousand population is simply not attainable and, again, local surveys are required to ascertain current demand. In the meantime, there is greater emphasis on indoor facilities, and surveys are needed of the use of sports centres, swimming baths and the whole range of urban sports.

Finally, rural community surveys are required to give the planner that necessary depth of understanding of situations in respect of which he is formulating proposals. Studies of population movement will be especially important, not only in areas which are experiencing net outflows of people, but also those districts which might now be gaining population on urban fringes by virtue of commuter settlement. Here, the nature of social change will be pronounced; accordingly studies of community integration and the use of, and demand for, different types of services will be required.

Additionally, the types of local rural surveys will be similar to the urban range listed above, with the exception that problems peculiar to large urban areas (particularly housing and the concentration of social problems) will not be represented. But there is the same need for local studies of villages or small towns, though without the contribution, because of scale, that Enumeration District Census data can give. The satisfactory provision of social services and community facilities in rural areas is a planning problem, and the rapidly changing social situation which has a bearing on this demands investigation: for example, the place of public transport, age structure and restrictions on mobility, and the critical population sizes of rural settlements. Lastly, we might mention the need for studies of the recreational patterns of a car-based urban population in rural areas; these will contribute to the satisfactory planning of country parks and of recreational facilities in the countryside generally.

These, then, are some of the fields in which social research can, and must, serve the town planner. For some of this research the planner

30

will be able to call upon the services of universities, consultants, market research firms and departments of the central government. But for much of it he will have to rely upon the resources and skills to be found in his own local authority department. Where the skills do not exist, they will increasingly need to be acquired. The remainder of this volume provides a detailed, practically-based outline of many of the techniques associated with these skills. It constitutes an introduction to social research techniques for the non-specialist engaged in research for planning.

Chapter 2

STAGES IN A RESEARCH PROJECT

There are several quite distinct stages in the development and conduct of a research project. Having decided on the objectives of the survey, or having been given the terms of reference, at least five major steps can be identified. First, consideration must be given to the method of data collection to be employed. Following this, decisions need to be taken about the kinds of data to be collected and the form in which they are to be collected. Where the basic method is to be a survey, a principal concern will be the design of the questionnaire. The fourth stage involves the drawing of a sample, so as to obtain data which are representative of the population that is being examined. Finally, there are the steps to be taken in data analysis.

Each of these stages is the subject of succeeding chapters, the principal issues being examined under the following headings: Self-administered and Interview Surveys, Questionnaire Design, Profile Data, Samples and Sampling Methods and Data Analysis. A further chapter deals with non-survey research techniques. In this way, the book provides a full review of the kinds of problems which an investigator faces in social research in planning. As an introduction to these chapters we will briefly indicate here the scope of the subject matter.

In Chapter 3, the advantages and disadvantages of self-administered and interview surveys are examined; also the situations in which they are best used. The problem of securing a satisfactory response rate is discussed, particularly from the point of view of the factors causing nonresponse and the means whereby the response might be increased. The section concludes with a review of the techniques of interviewing and the general problem of bias in interview and self-administered surveys.

Chapter 4 deals with questionnaire design. This looks at the decisions to be made before the questions are drafted, the type of questions to be asked – for example, whether they are to be 'open' or 'closed' – and the actual wording of the questions themselves.

The importance of pilot work in aiding design is particularly stressed.

Profile data is the subject of Chapter 5. The problem examined here concerns the kinds of data to be collected and the ways these can be grouped for easy analysis. Factors concerning a number of characteristics of the survey population are considered: age, sex, marital status, family or household composition, education, income and occupation (or social class).

Chapter 6 is concerned with the question of samples, sampling frames and sampling methods. With regard to the determination of the size of sample, an outline is given of the use of the formula for the calculation of standard error.

In Chapter 7 some of the general equipment available for data analysis is described. Steps to be taken in data analysis are outlined and suggestions made about the organization of the work of analysis.

Finally in Chapter 8, four non-survey research techniques are described and suggestions made as to their applicability. These are the use of physical evidence, mechanical and electronic devices, documentary sources and observation.

It is not the intention of this volume to be a manual describing how different types of social research might be undertaken. It is, rather, a discussion of the kinds of problems the researcher faces in social investigations which are relevant for the planner. We have therefore concentrated, first, on identifying the different stages which usually structure particular projects and, second, on examining the different techniques which might be employed at each stage. Thus, the volume is, essentially, a guide to the range of relevant techniques that are available for social research in planning.

Research projects differ by virtue of their subject matter and objectives, and the survey techniques adopted will vary accordingly. But the stages through which projects pass are similar and it is important to impress upon research workers the need for disciplined thinking around a particular project within this fairly regular framework. To illustrate the basic universal nature of these stages two very different projects, with which the authors have been connected, will be briefly described, ranging from presentation of objectives to data analysis; the first is the *Birmingham Recreation Planning Study*,[65] the second, a survey of population migration in the Northern Region.[46]

The Birmingham Recreation Planning Study

The purpose of this study was to test various possible ways of acquiring data about recreation demands which would be meaningful for planning. This objective needed more precise formulation and four specific separate aims were proposed:

I. to discover whether, by the joint use of interviews and self-administered questionnaires, a significant increase above the norm could be obtained in the number of respondents to a survey without a correspondingly significant increase in costs and a decrease in the quality of data obtained;

II. to test the validity, and to assess the potential use, of the concept of *recreation types* as an aid to planning;

III. to seek to provide guidelines for future studies relating to the collection and analysis of profile data;

IV. to investigate, in a general way, the interrelationships between recreation supply and demand.

The first stage in research that we have identified concerns the choice of methodology to be employed. This involves consideration of the kinds of data required, their sources and the procedures to be followed in collecting and analysing them. The objectives of the Birmingham study dictated the use of a number of different methods and the final decision was to employ household questionnaires by interview, and recreation activities questionnaires by interview or self-completion. It should be noted that these procedures involved the hiring of 'outside' interviewers by contract.

The next stages concern the actual collection of these data, involving a number of separate steps. In the Birmingham study these were:

I. the design of the questionnaire: because analysis was to be by computer, the design had to suit coders and punchers; because collection was by interview and self-completion, the design had to suit both interviewers and respondents.

II. the profile data to be collected: because this was a major objective of the study, questions were necessary over a wide range of profile characteristics.

III. pilot work: the primary aim here was to test the design of the questionnaire.

IV. sampling: a systematic 1-in-8 sample was drawn from each street in the survey area.

The final stage, the analysis of the data and the preparation of the report, involves several considerations. The analysis of data can be examined at two distinct levels – the operational (involving the use of

34

machinery) and interpretive or statistical. Because an electronic computer was used in the Birmingham study, coding and punching had to be done. Certain interpretive analyses were then carried out on the computer, such as a factor analysis and statistical tests to validate the profile data. In the writing of the report it is necessary to give the answers to questions posed in the objectives of the study (or at least to show that answers have not been found). In the Birmingham project the report was fully comprehensive as to the aspects of work covered and presented as an academic report suitable for publication.

The Study of Population Migration

A comparison with the study of migration in the Northern Region reveals the similarity of overall framework between the two studies, but emphasizes internal differences dictated by objectives and constraints.

This survey was concerned with the people who were moving within, into and out of the urban parts of the Northern Region:

    I. who the migrants were: their age, sex, family composition and occupational status;

    II. where the migrants went and what were the main streams of movement;

    III. what were the reasons for migration, and how far did these differ between people of different characteristics.

Data were required from a sample of migrant households who had actually moved house over a given period of time (1965-6). These data were required at sub-regional level (ten sub-regions were identified) as well as for the Region, and therefore, to be statistically viable, a large number of persons had to be contacted over a wide area from West Cumberland to Durham and from Northumberland to the North Riding. Limited financial resources were available, and this seemed to preclude any use of 'outside' interviewers on a large scale. These factors dictated the use of a postal questionnaire for data collection.

The steps in the collection of data can be examined in the same manner as for the Birmingham study:

    I. design of the questionnaire: major constraints were the need for computer analysis and self-completion by respondents. Additionally, the importance of the impact of a suitable, accompanying letter had to be borne in mind. Every effort was made to draw up a visually attractive, short questionnaire and to communicate to the respondent the significance of his own contribution in the overall volume of replies.

II. the profile data to be collected: these were restricted to the needs of the study, namely to identify who the migrants were in terms of age, sex, social class and housing tenure.

III. pilot work: four types of questionnaire were tested, varying between types of questions asked and the order in which they appeared. The postal questionnaire method was also tested for response rate.

IV. sampling: migrants were identified largely from a comparison of Electoral Registers relating to the qualifying dates (October 10, 1964 and 1965). In practice this did not necessitate a direct comparison of the two Registers, but could be done by using a draft list of additions to the Register. A check had to be made on the sampling frame to discover whether the source for identifying the migrants was sufficiently accurate and reliable.

With regard to the analysis of data, principal steps were:

I. coding, carried out on specially designed transfer sheets for which detailed instructions were prepared for the coders;

II. weighting, to modify crude figures by relating them to a common base;

III. tests for bias, stemming from non-comprehensiveness of the sampling frames and from various kinds of nonresponse;

IV. tabulation design;

V. punching of coded data on to eighty-column punched cards;

VI. operational analysis of data on the computer;

VII. interpretive analysis of data as part of the writing of the report.

Finally, there was the writing of the report itself and the presentation of the findings. This project was a local authority enterprise in which a number of different planning authorities had pooled financial and manpower resources and had worked closely with certain Government Departments in the Region. A number of different clients had therefore to be satisfied in the one Report and various levels of reader were involved from the elected member to planning officer and specialist research worker. Three volumes of the Report were therefore published, one a short and easily-read summary, the second a detailed report of findings and implications, and the third a section on research method.

Comparison of these two projects shows, therefore, a broad similarity as far as structural framework is concerned, in that there are the same stages which the researcher should recognize, and this framework extends, of course, to a large number of projects. Within each of the stages of the structure, however, there are a variety of factors to be considered in deciding the techniques to be employed; hence, not only do the two studies which have been outlined show significant differences at this level, but other studies will also vary. The following chapters describe the alternative research methods, their weaknesses and their strengths, which might be adopted.

Chapter 3

SELF-ADMINISTERED AND INTERVIEW SURVEYS

Considerable controversy currently surrounds the use of self-administered surveys in social research. These are surveys in which questionnaires are completed by respondents themselves; there is no interviewer. The most usual form of self-administered survey is the postal or mail survey, but there have been cases where the questionnaires have been delivered to respondents by other means – for example, by handing out questionnaires at car parks. Some investigators extol the advantages that self-administered surveys have over interview surveys in terms of cost, convenience and comparability; others point to their one great disadvantage – the relatively low rates of response that are frequent in such surveys, which can often lead to significant bias in the findings. There is no doubt that the problem of nonresponse is the most difficult one to overcome; but it is not insuperable, and if it is overcome, there is much to commend their use.

A main advantage of the self-administered survey as a research technique is that it is possible to cover a wider geographical area and to reach a larger sample of the population (with given financial resources) than is possible by the use of an interview survey. Alternatively, the costs of reaching a given sample of the population will usually be lower for a self-administered survey than for an interview survey. Furthermore, generally, the degree of organization required for a self-administered survey is less complex than for an equivalent interview survey or observational survey; thus, for example, the selection, training and instruction of interviewers and observers is not required. It is also argued that the respondent will answer a self-administered questionnaire more frankly than he would an interview, since anonymity is not only assured but is seen to be assured. Again, it is claimed that the questions in a self-administered survey are totally standardized and, therefore, responses are totally comparable. Other merits are that it avoids the difficulties arising

38

from respondents' antagonism towards interviewers; that it can be answered at the convenience of the respondent, rather than at the convenience of the interviewer; that it usually secures a greater response from male members of the household than does the interview survey; and that it is often easier to reach mobile sections of the population (by postal survey) than by interview – for example, people who work unusual hours or who are away from home for several days (or more) at a time.

Against these advantages there is the great disadvantage that response rates to self-administered surveys – and, in particular, to postal surveys – are often very low. Experience suggests that a return of between 30 and 50 per cent is usual, and that a return of above 50 per cent is good. Moreover, the people who do return questionnaires may not be representative of those to whom they are sent; that is, the sample may be distorted significantly by the degree of nonresponse. Further complications are that the respondent may misread or misinterpret questions, and there is no interviewer available to correct him; that unsatisfactory or incomplete questionnaires cannot easily be returned to respondents for correction; and that a self-administered survey must usually be spread over a much longer period of time than an equivalent interview survey, since the investigator has no control over the timing of returns. Another disadvantage is the bulkiness and possibly repellent nature of a questionnaire received through the post – the number of questions that can be asked is, therefore, limited.

Clearly, the vital limitation of self-administered surveys lies in the difficulties of obtaining an adequate response, and, hence, the possibilities of creating bias in the findings. There are two elements of concern here; firstly, to examine methods of increasing response rates; and secondly, to investigate methods of detecting and measuring the significance of the bias arising from nonresponse.

The initial response to a self-administered survey depends upon such factors as the characteristics of the population to whom the questionnaire has been sent, the degree of interest in the subject of the survey that can be aroused in this population, the standing and prestige which the sponsor of the survey has amongst his population and the effect which the covering letter engenders among the recipients. There is no very substantial guide concerning these factors, although a number of studies have produced some indications. Sletto[57] has indicated that the length of the questionnaire does not seem to have a significant effect upon response rates (although, in fact, he only tested questionnaires of ten pages in length, twenty-five pages in length and thirty-five pages in length. He believes that there might be a pronounced difference in response rates as between questionnaires of one and ten pages in length). Sletto's work also showed that an altruistic appeal to recipients is marginally better than a challenging one; and that a postcard follow-up is usually as effective as a letter. Seitz[55] reported that the use of postage stamps on reply envelopes is more effective than reply-paid envelopes. Clausen and Ford[11] suggest that the covering letter is best separated from the questionnaire itself.

Another method of attempting to increase the size of the initial response is to offer some inducement to respondents. Watson[69] claims that financial incentives are relatively ineffective. On the other hand, Shuttleworth[56] and Hancock[30] both found that the inclusion of a 25-cent coin with the questionnaire appeared to have a significant effect upon rates of response. In the former case, the response rate was 52 per cent for those persons who were sent a coin with the questionnaire as compared with 19 per cent for those who were sent the questionnaire only. In the latter case, the porportions were 47 per cent and 10 per cent respectively. A similar kind of inducement is to offer respondents the opportunity of winning a prize of some kind. A theatre survey in Newcastle upon Tyne in 1967–8 placed the numbers of completed questionnaires into a lottery in which a prize of free tickets for a future performance at the theatre was to be offered.[44] There do not appear to have been any systematic studies of the effectiveness of prize offers, but *prima facie*, it seems to offer distinct possibilities. (We are told that the British are a nation of gamblers at heart!)

Distinct from the methods of increasing the size of the initial response are those which aim at securing additional responses. The most well-known of these is the follow-up letter, or reminder. The general evidence seems to be that follow-up letters *do* bring in additional responses; but there is some doubt about the extent to which they are effective. Gray and Corlett[25] found that the total response to one survey was almost doubled as a result of the follow-up letter – from 38 per cent to 70 per cent. Other investigators,[42] have also found that a follow-up letter brings in additional responses, although not in such dramatic proportions. In particular, many investigators have found that the greater the number of follow-up letters (i.e. a second, third, and fourth, etc.), the higher the proportion of responses that are likely to be obtained – within certain limits, of course. Evidence from the survey of population migration in the Northern Region[46] in 1967 showed that the response rate tailed off after about ten days and a reminder letter was instrumental in achieving an ultimate response of nearly 70 per cent.

After efforts have been made to increase response rates there still remains the problem of how to determine the bias that has arisen from the exclusion of the nonrespondents from the sample. A number of methods have been suggested for dealing with this problem. One suggestion has been[42] that interviewers be sent to nonrespondents with the purpose of persuading them to complete the questionnaire and so become respondents; or, at least, to obtain some data about their characteristics so that these may be compared with the characteristics of the respondents. Another suggestion has been that respondents who return their questionnaires only as a result of follow-up letters are more akin to nonrespondents in their characteristics than to respondents. Hence, the characteristics of nonrespondents, and thereby the resulting bias in the sample can be deduced from the characteristics of follow-up respondents. There is a relative lack of systematic evidence to support this assumption, but a few studies[61] have tended to corroborate it rather than to disprove it. A third method of assessing the characteristics of non-respondents is by reference to records, wherever this is possible. In a study relating to previous students of the State College of Washington, Reuss[53] had access to college records. Thus, he was able to assess the characteristics of nonrespondents in considerable detail. He found that 'the intelligence of the questionnaire recipient, his qualities of purposefulness and initiative, his loyalty or strength of ties attaching him to the institution or individual sponsoring the questionnaire and a rural background seem to be factors favourably influencing questionnaire response'. This type of empirical study might usefully be undertaken in Britain. It establishes clearly that there is a significant difference in the characteristics of respondents and nonrespondents, although Reuss does not indicate whether the respondents to follow-up letters are most like nonrespondents or the initial respondents.

INTERVIEW SURVEYS

There are many advantages in the use of the interview survey as a research method. The greatest of these is, of course, its flexibility. The skilled interviewer can make sure that the respondent fully understands the nature of the information that he or she is being asked to give; he, or more usually, she can probe more deeply into the subject's responses; she can show the respondent cards, lists and similar material and so focus his attention more completely on the subject of interest; above all she can establish rapport with the respondent, and thereby, maintain the latter's interest and participation in the survey.

Other advantages are that interview surveys generally yield a high response, either because people are usually willing to cooperate when approached personally, or for the negative reason that they do not wish to appear un-cooperative; that they can be designed to yield an almost perfect sample of the general population more easily than any other research technique; that the information secured through the interview survey is likely to be more accurate than that secured by any other technique, since the interviewer can make sure that the respondent fully understands the meaning of questions; that the interviewer can obtain supplementary information about the respondent which might be withheld or distorted by other methods – for example, profile data; that the interviewer can present visual material to the respondent; that the interviewer can usually control which person, or persons, answers the questionnaire; that a skilled interviewer can handle subjects about which the respondent is likely to be sensitive much better than can be done by any other technique; and that the language of the survey can be adapted by the interviewer to suit the ability and/or educational standard of each respondent.

There is, however, a number of disadvantages, some of which spring directly from the above advantages. The most important is the problem of over-rapport, leading to bias. The interviewer may be so successful in establishing rapport with the respondent that she unconsciously affects the latter's responses. She may unwittingly influence the respondent's replies by such minor details as her appearance and tone of voice; but, if she has established a high degree of rapport, her influence may become more serious. The respondent may tend, for example, to give answers and express opinions which he thinks the interviewer expects to hear from him or which he thinks the interviewer would approve of. Another disadvantage springs from the personalities and characteristics of the interviewers themselves. The interviewers' own expectations and their selective understanding and recording of responses may produce bias; interviewers react differently, for example, to different respondents. Again, they may probe responses with more or less care, depending upon their (unconscious or subconscious) assessments of the different respondents. These and similar characteristics can easily produce bias in the survey data.

Other problems are that the costs of transportation, time, organization, and so on, may be such as to make the interview survey impracticable – or they may be disproportionate to the additional data that can be obtained by using the interview survey rather than a different technique, such as the postal survey; that interviewer bias

42

may distort the data that are obtained – that is, the interviewer may record the information that she expects to hear from a particular type of respondent rather than that which she actually hears (this distortion can be deliberate or, more usually, unconscious); that the degree of organization involved in the interview survey is more complex than for most other research techniques – for example, the selection, training and supervision of interviewers; and that it may be difficult to locate respondents when these are to be very specific – for example, in household surveys, it may be difficult to make contact with people who work unusual hours or those whose work takes them away from home for several days at a time.

This outline of the advantages and disadvantages of interview surveys refers to these as a homogeneous group. There are, however, two broad categories of interview surveys, grouped according to:

I. the interview technique that is used – *standardized, semi-standardized* or *non-standardized*;
II. the location of the interview – *in the household*, or *on 'site'* (by on 'site' is meant at the place where the subject matter of the survey generally occurs – for example, a shopping centre or public park). Site surveys may relate to either the *users* of facilities and services, or to the *suppliers*, or to both.

These two categories are, of course, not mutually exclusive.

The *standardized* (or structured) interview is used when the same (or similar) information is to be collected from each respondent. This means that the answers of all respondents must be comparable and classifiable; differences in responses must reflect actual differences among respondents, and not differences arising from the different questions asked or the differences in meaning which respondents have attributed to the same questions. Some writers have taken this to its extreme conclusion: 'Only if all respondents are asked exactly the same questions in the same order can one be sure that all the answers relate to the same thing and are strictly comparable. Then, and then only, is one justified in combining the results into statistical aggregates.'[42] This is an extreme view, and there are situations in which question ordering can be alternated without destroying validity and the reliability of responses. The standardized interview is, however, based upon this concept. The wording and sequence of questions are determined in advance and these are, in principle, asked of all respondents in exactly the same way.

The use of the standardized interview is based upon three import-ant assumptions: first, that respondents have a sufficiently common

43

vocabulary so that it is possible to formulate questions which have the same meaning for each of them; second, that a uniform wording for all respondents can be found for any subject matter; and third, that if the meaning of each question is to be identical, and since all preceding questions constitute part of the context, the sequence of the questions must be identical. There is also the implicit assumption that the requirements of these three separate assumptions can be met – by means of piloting and pretesting of questions and questionnaires.

These assumptions are very sweeping and can be seriously challenged. The assumption about a common vocabulary is more or less true in relation to the degree of homogeneity among respondents – their backgrounds, experience, personal characteristics and so on. The more heterogeneous the respondents the less likely it is that they will have a sufficiently common vocabulary. Of course, the investigator can try to introduce a homogeneous element by such methods as phrasing the questions in the form of the *lowest common denominator* or by the use of *filter* questions. The second assumption is more realistic, but still raises some difficulties. The development of a special language or *jargon* for certain subjects can cause awkward problems; the uninitiated will not understand the jargon and must, therefore, be asked questions in informal terminology; the initiated may assume that the informal terminology means something different from the jargon since, otherwise, he would expect to be asked the questions in the jargon. The third assumption is the least satisfactory; it means that the emotive impact and 'sensitivity' of each question are assumed to be identical for all respondents. This raises particular difficulties in the case of opinion or attitude questions as compared with factual ones.

Broadly, then, the standardized interview is most helpful in situations requiring the collection of data from groups which are significantly homogeneous; about subjects which can be expressed fairly easily in non-specific language; and covering matters which are factual or relatively non-emotive. In planning research such situations would be covered by factual surveys about such matters as the use of parks and other public facilities, housing types and amenities, origin-and-destination surveys, and so on: in fact any subject which does not require too much questioning about attitudes.

Difficulties of the kind outlined above have led to a greater interest in and use of the *semi-standardized* interview. The advocates of this method reject the idea that comparability and, hence, statistical aggregation can only be achieved by the use of identically-worded questions and identical sequence of questions for all respondents.

Rather, they argue that, if questions are to have the same meanings for each respondent, they must be formulated in wording that is appropriate for each respondent. Instead of operating with a pre-determined schedule of questions, the semi-standardized interview is based upon a list of required information which is given to the interviewer. In its simplest form the semi-standardized interview utilizes a schedule or questionnaire which the interviewer may present to the respondent in any sequence of questions. At its most refined level, the interviewer operates only with a list of the data which she must obtain.

The basic assumptions underlying the use of the semi-standardized interview are also threefold: first, if the meaning of a question is to be standardized and, hence, the responses made comparable, the question must be formulated in words familiar to and commonly used by each respondent; second, no fixed sequence of questions is satisfactory to all respondents – the most effective sequence for any respondent is determined by his readiness and willingness to take up a topic as it comes up; and third, through the stringent selection and training of interviewers, the necessary skills can be obtained by which the question-wording and sequence can be designed so as to give equivalence of meaning for all respondents.

An argument against the first of these assumptions is that different question wordings sometimes produce markedly different responses. This has been demonstrated by a number of investigators.[7] The second assumption challenges directly the third assumption about standardized interviews – namely, that the sequence of questions must be the same for all respondents in order that the context is identical and, hence, the meaning. The most telling argument here is, of course, that identical sequence does not necessarily imply identical meaning. It is, however, the third assumption about semi-standardized interviews which is the most difficult to sustain. It is likely that no amount of rigid selection and skilled training of interviewers will produce one who can give perfect equivalence of meaning to all questions to each respondent. The real problem, there-fore, is whether or not errors arising from differences in meaning can be reduced to a level at which they do not significantly affect the research findings, and not whether these errors can be totally eliminated.

The semi-standardized interview appears to be most useful in situations requiring the collection of data from groups which are heterogeneous (as defined by some characteristics which the investi-gator assumes, or discovers by pretesting, to be significant); and which cover matters which are emotive or sensitive. In planning, such subjects as opinions about major planning proposals, attitudes

45

to comprehensive redevelopment, or reactions to living in high rise buildings, might be relevant.

Standardized and semi-standardized interviews are, basically, the required techniques for situations in which the same categories of information are sought from all respondents. The objectives of the *non-standardized* interview are quite contrary to this. The user of the non-standardized interview makes no attempt to obtain the same categories of information from each respondent and there is no necessity for the unit of analysis to be the individual (although it usually is). The method may be used to explore a broad problem, such as the social effects of technological change, or the socio-psychological effects of a university education upon children from an unskilled working class background. The essence of the non-standardized interview is that its content (i.e. questions) can be varied from one respondent to another. Hence, there is no predetermined schedule or questionnaire and, often, no check-list of required information. Usually, there is no predetermined population sample for study. At its most formal, it approaches the semi-standardized interview. This is called the guided or *focused interview*, which covers a predetermined set of topics but allows the interviewer total freedom in the methods he uses to obtain data relating to these topics. The respondent is given the opportunity to develop his experiences and attitudes in his own way and at any length. The interviewer is required to guide the interview in such a way that all the predetermined topics are covered in the course of the interview. The objective of the focused interview is to centre attention upon a particular set of topics (often experiences) about which the interviewer wishes to know the respondent's participation or views. The respondent is asked to look back in order that the interviewer can discover his responses in the original situation to which the interview refers, rather than to ascertain his general feelings about the situation as they are now. At its most informal, the non-standardized interview consists of conversation and entirely non-directive interviewing. It has been likened in this respect to the psychoanalyst's couch! There is no predetermined set of topics to be covered and the approach is entirely free.

All forms of the non-standardized interview require highly skilled interviewers. The interviewer has very great freedom in the formulation of the content of the interview and in the questioning procedures. The questioning will, therefore, stem from the interviewer's understanding of the overall objectives of the inquiry. Hence, non-standardized interviews will generally be conducted by the researchers themselves.

It is worth noting that the clear distinction drawn between standardized, semi-standardized, and non-standardized interviews in the above discussion does not indicate a rigid distinction in their use for specific studies. There is no reason why two (or, even all three) methods cannot be used in a single study. Indeed, it is possible to construct a schedule or questionnaire which is partly standardized in its form, but which includes a list of items of information which the interviewer is free to pursue by semi-standardized interview.

Interview surveys in the *household* may involve only one individual in each household or all members of the household (or, indeed, certain members of the household only – for example, those between the ages of 15 and 24 years). Household surveys may be on any scale from a single neighbourhood to a national inquiry. *Site* surveys may be made of the users or the suppliers of facilities and services. *User* surveys yield information about the users of a particular facility, particularly the intensity of use. The main disadvantage is the in-built restriction that non-users are automatically excluded from the study. User studies may be made by interviews with individuals and groups of individuals, or by reference to organized clubs and other bodies. The major limitation of the latter lies, however, in the fact that large numbers of users of certain facilities may not be members of clubs and organizations. Site surveys of *suppliers* may be made by reference to the observation of facilities actually available, or by interviews with the persons responsible for the provision of facilities.

A general problem in all survey research is the problem of non-response. It is a problem of particular and crucial significance in postal surveys, where rates of response are, generally, relatively low and hence can easily lead to sampling bias. In interview surveys, the problem is primarily an organizational one, closely associated with the selection, training and instruction of interviewers. It consists of two quite distinct elements:

I. *non-contact*, arising from such factors as the removal to a new address of the individual or family whom the interviewer has been instructed to interview;

II. *refusals* to participate in the survey by those people with whom the interviewer has managed to establish contact.

The first of these elements is entirely random and outside the control of both the investigator and the interviewer. All that the investigator can do is to make sure that his sampling frame is as accurate and up to date as possible. The second element is, however,

47

susceptible to some degree of analysis and control by the investigator.

There is a relative dearth of research into the causes of interview refusals and the characteristics of refusal-prone interviewers. One investigation has been made, however, the main findings of which were that proportionately to their numbers, interviewers with high refusal-rates did about as many interviews as other interviewers; that high refusal-rate interviewers had had, on average, a substantially shorter length of service than low refusal-rate interviewers; that contacts in the higher income groups were more likely to refuse an interview than those in the lower income groups; and that, because interviewers have been trained in the tradition that to obtain a high-refusal rate is a sign of incompetence, there is likely to be some attempt, conscious or unconscious, at concealment of refusals.[23] These findings have certain implications for the selection, training and instruction of interviewers. It appears that the numbers of interviews carried out by each interviewer during a particular survey do not significantly affect refusal rates; but, on the other hand, experience – measured in terms of the length of service of the interviewer with the survey organization – does appear to have an effect. This suggests, *prima facie*, that experienced interviewers should, perhaps, be given relatively higher interview quotas than their less-experienced colleagues. Again, it is clear that refusals result, at least in part, from the characteristics of the contacts (or more precisely in this case, from one such characteristic – income levels). No satisfactory method for combating this problem has yet been suggested.

Several writers have developed criteria upon which the selection of interviewers should be based. These range, in content, from an ability to talk freely with all types of people to good health and plenty of physical energy. The most comprehensive set of criteria were outlined by Parten[49] in 1950, and, despite the passage of time, this appears to be the best statement. The criteria are best expressed in the form of characteristics that should be possessed by the ideal interviewer. The most important of these are as follows:

Characteristics which relate to the Interviewer's personality:
    I. An ability to talk easily with all types of people.
  *II. An ability to judge people and situations quickly and correctly and, allied to this, persistence and thoroughness; also quick wit and resourcefulness.
    III. A sympathetic as well as an enthusiastic interest in people.
  *IV. Conscientiousness, honesty and reliability.
    V. An inquiring mind.

48

*vi. An appearance and manner which inspires confidence.
vii. Good and reliable health, and plenty of physical (and mental) energy.

Characteristics which relate to the Interviewer's competence:
  *i. Keen powers of observation and a regard for detail.
  ii. A good memory (and where possible an ability to take shorthand).
  iii. Legible handwriting.
*iv. An ability to summarize and record objectively; in particular, a freedom from bias in observation and in the eliciting and recording of facts and opinions.
*v. Freedom to travel, if necessary, and willingness and ability to work unusual hours.
  vi. Preferably an education to at least Ordinary Level of the GCE.
  vii. It is often also useful if the interviewer can drive and has the use of a car.

It would be unwise to attempt to rank these criteria in an order of preference – if only because the desirable characteristics will vary from one survey to the next, depending upon such factors as its subject matter, the special characteristics of the respondents, and so on. There are, however, some criteria which would appear to be fundamental to all interview surveys and these have been marked with an asterisk in the preceding list.

It is difficult to assess the value to be placed upon each of these various characteristics when selecting interviewers. This is particularly so since the possession of these characteristics is not entirely fortuitous. Some of them can be implanted in the interviewer during his or her training. In other words, the selection of interviewers cannot be entirely divorced from their training. The training should be part of the selection procedure and interviewers should be made aware of the fact that selection is provisional until they have completed part, at least, of their training.

There would appear to be eight major steps in the training of interviewers. First, they should be given a sound general briefing on the nature and objectives of interview surveys, and of the qualities that will be required of them. This briefing should not be too technical, but the meaning of technical terms and jargon must be made absolutely clear. They should then be introduced to the specific tools of the survey – questionnaires, check-lists, interviewer report forms and so on – and required to show their understanding of these by taking written tests which pose hypothetical interview situations.

D

These written tests can then be followed by demonstration and trial interviews with volunteer-subjects. The fourth stage is for trial interviews to be undertaken in the field. Interviewers should then be given information about, and some experience of, the coding stage of the survey, since this is the one which immediately follows their own work. The training can continue when the interviewer is put on to her first actual surveys – by making detailed checks of her early returns. This process can be continued by 'spot' supervision and observation of the interviewer at work. Finally, continuous training can be provided by random tests of the interviewer during field work.

The main conclusions that emerge from this outline are twofold: firstly, interviewers should be thoroughly conversant with the techniques, difficulties and problems inherent in their work; secondly, training should be continuous throughout the interviewer's working life and not a procedure which lasts for a few weeks at the beginning of her career. In many local planning departments all of these desirable features will not be achieved, because many short surveys will be carried out by staff and supporting students on an irregular basis. But at least the *nature* of the briefing is shown to be important: it is vital that the interviewers know precisely what is expected of them.

In addition to training in the techniques of interviewing, the interviewer will, of course, require specific instructions relating to each individual survey which she undertakes. The Government Social Survey has produced a general code of practice[4] which should prove invaluable; but, in addition, there are certain basic matters about which the interviewer should receive clear instructions. She should be thoroughly briefed on her instructions for the survey and should also be kept up to date on those instructions. She should be informed clearly of the chain of responsibility for the survey; that is, she should know exactly whom she must contact when in need (including the names of deputies for this person). She should not be permitted to delegate or subcontract any of her interviewing quota, nor to combine any other activity (for example, door-to-door sales) with her interviewing. She should be provided with a letter or certificate or badge of authority. And she should be given clear information concerning the method, frequency and procedure for her payment. In addition to these general matters, she should, of course, be given specific instructions concerning the questionnaire; for example, sequence of questions, use of prompt cards, whether or not she may probe for responses to particular questions, and so on.

The various criteria for selection of interviewers, suggestions for training and proposals for instruction that have been outlined above refer to a situation in which the research investigator has a high degree of control over the conduct of field work. Many research organizations, however, subcontract the field work for their research to market research organizations. In these circumstances the degree of control is significantly reduced. A great deal will then hinge upon the choice of subcontractor. At the least, no choice should be made until the investigator has visited the organization's field control centre and learned something of its administrative structure and capabilities. This is of the utmost importance, for so much hinges upon the abilities of the interviewers. The results of a survey may be significantly affected by the work of the interviewers; for, one of the major disadvantages of interview surveys – to which attention has been repeatedly drawn in research writings – is the possibility of bias arising from the characteristics of the interviewers themselves. The interviewers' own expectations and their selective understanding and recording of responses may produce bias. For analytical purposes, we can distinguish three kinds of interviewer bias – that arising from the interviewers' personal characteristics, that arising from their opinions, and that arising from their expectations.

The predominant concern of many research investigators has been, until recently, with the possibilities of bias arising from the personal characteristics of interviewers, and their opinions. It has been argued that such personal characteristics as the sex, age and social background of the interviewer can influence the answers obtained from respondents. Thus, for example, Wilkins[75] found, during his inquiry into the demand for campaign medals after the Second World War, that 'the ex-servicemen gave replies to elderly women interviewers which showed a greater desire for medals than the replies they gave to young women'. Similarly, it has been argued that strongly-held opinions of interviewers would be unwittingly communicated to respondents – by such means as the manner in which questions are asked – and would, thereby, influence the answers obtained from the respondents.

Studies by the National Opinion Research Centre in the United States of America[31] have suggested, however, that the biasing effects arising from the personal characteristics of interviewers, and their opinions and ideology, are much less significant than is commonly supposed. These studies suggested that a much more significant source of bias is the expectations of interviewers. These expectations-biases are of three main kinds:

I. *attitude-structure expectations*: responses to the early questions during an interview may give the interviewer an indication of the respondent's attitudes: she may then expect consistency

51

from the respondent and interpret later responses, especially marginal or ambiguous ones, in the light of her conception of the respondent's attitudes;

II. *role expectations*: the interviewer may be given an early impression that the respondent is of a particular (precon-cieved) type; she may then record later responses, especially ambiguous ones, in the light of the responses she would expect from this type of person;

III. *probability expectations*: the interviewer may expect a certain distribution of attitudes, opinions and characteristics among the survey population, and may interpret ambiguous and doubtful responses in accordance with this expectation.

The American studies could find very little positive evidence to support the existence of bias arising from probability expectations, but they found clear evidence relating to the biasing of attitude-structure expectations and role expectations. One study[58] indicated that interviewers often record the response that they expect to hear from a particular respondent. Discussions with interviewers showed that many of them believed that the attitudes of any particular respondent are bound together in an organized structure; and, on the basis of this belief, they expected respondents to answer questions in a consistent manner. By means of phonograph transcriptions, a group of interviewers heard two typical, yet markedly contrasting, respondents to an interview situation. The interviewers were requested to record the responses of these respondents on copies of the questionnaire with which they had been supplied. The results clearly indicated that the interviewers' recording of responses was affected by attitude-structure expectations. In particular they suggested that the inexperienced interviewer is less likely to feel pangs of conscience, or may be less aware of the strict rules governing coding and is, therefore, more prone to expectation effects.

Given that bias arising from interviewer characteristics, opinions and expectations does exist, there are two courses of action open to the investigator: to attempt to eliminate it and, to attempt to measure its significance. The chief method of elimination would be to improve methods of selection and training of interviewers. Detection and measurement of bias may be obtained in a number of ways. For example, the data obtained by the interviewer, or part of it, may be capable of verification by recourse to records – such as birth certificates and employer's records. Again, the interview data may be subjected to consistency checks, by such methods as asking for the same information from two different respondents (husband and wife) or by asking the same person for the same information in two different ways. This type of check is, however, negative in concept, since if the two sets of responses differ, there is no way of knowing which is correct. All that is available is a measure of variation. A third method of checking is to re-interview respondents with a different interviewer. Again, however, the method is negative in concept, since there is no way of knowing which of the two different sets of responses is correct. Moreover, there is the additional problem that the respondent's answers may actually differ in the two interviews (respondent's bias).

One particular type of interviewer bias is that arising from the deliberate falsification of responses by the interviewer. It has already been noted, that, because interviewers have been trained in the tradition that to obtain a high refusal rate is a sign of incompetence, there is likely to be some attempt, conscious or unconscious, to conceal refusals. One method of concealment is to fabricate responses. Wholesale interviewer cheating is not, usually, difficult to detect. Most survey organizations employ postal checks on a sample basis. This is a well-tried research technique. There have been, however, some interesting extensions of the postal check.

Most checks are simply concerned to inquire of respondents whether or not they were in fact interviewed during a particular survey. Respondents are asked only whether an interview was carried out with them and whether or not they found it very interesting. The response rates to these postal checks are often relatively low. Gray and Harris,[26] however, carried out a series of postal checks in which respondents were asked to confirm or supplement data obtained from the interviews. The response rates to these postal checks seemed above average; and the authors have suggested that 'informants treat our letters more seriously just because we are asking them to confirm or supplement information given during the interview'.

Postal checks are usually adequate to detect wholesale interviewer cheating. It is much more difficult, however, to detect cheating on a subtler scale – that is, the fabrication of responses to particular questions or parts of questionnaires. In this case, 'prevention rather than cure' seems to be the most satisfactory approach. In other words, attempts should be made to weed out interviewers who are prone to cheating, by the selection and training procedures outlined earlier. In particular, detailed checks on interviewers' returns and periodic (random) tests during field work may be employed.

In addition to interviewer bias, there is, of course, the possibility of bias arising from incorrect response on the part of the respondents. The latter may give incorrect answers for various reasons: they may lack the necessary knowledge; they may misunderstand or misinterpret questions; they may, consciously or unconsciously, wish to withhold the true information. This type of bias is very difficult to detect. One method, is, of course, to ask for the same information in two different ways during the course of the interview. In addition, the well-trained interviewer may be able to throw doubt upon the respondent's answers by simply recording her observations during the course of the interview. Again, some of the responses may be capable of being checked by reference to records of one kind or another. In the long run, however, the validity of responses is almost inevitably dependent upon the (presumed) fact that the majority of people are truthful.

It is, perhaps, worth making one final point about interview surveys, relating to the nature of respondents. An advantage of such surveys is that the investigator has, in effect, a captive audience – for example, on the doorstep or in a countryside car park. There are, however, certain disadvantages which spring from this characteristic: for example, distraction of the interviewer, or where young children are present, of the parent (who may be the respondent). An instance of such disadvantages occurred in a survey of urban parks in Newcastle upon Tyne in 1967;[45] one interviewer had her coat set on fire, and another had her handbag stolen!

There are certain general guidelines for planners engaged in social research which spring from this discussion. Essentially, the planner who is directing an interview or postal survey must be aware of the inherent weaknesses of the method which he is using. It will never be possible to eliminate all weaknesses, but this does not matter. What is important is that he should recognize the potential for bias which springs from weaknesses of method. Thus, in postal surveys, emphasis must be placed upon any and all techniques which might increase response rates. In interview surveys – especially when inexperienced department staff and student interviewers are to be used – it is vital that full and detailed instructions are given about the conduct of the interview. In particular, the interviewer must know to whom he or she can turn when in difficulties. The importance of obtaining unbiased information should be stressed and, hence, particular instructions given about when the interviewer can and cannot probe for a response. If these simple rules are adhered to, there is every chance that the survey will be successful. But there it one golden rule above all: in any case where the investigator cannot resolve to his own satisfaction whether or not a particular interview schedule reflects interviewer bias it should be rejected, and eliminated from the returns.

Chapter 4

QUESTIONNAIRE DESIGN

The design of a questionnaire is intimately related to the general plan of the survey. After the planning stage, some specification for a questionnaire can be drawn up, which will follow directly from the statement of the issues to be investigated and the overall design adopted for the survey. The objectives of the survey should be absolutely clear before the design of the questionnaire is attempted. In the discussion of methods and techniques which follows, it is suggested that there is no method and no system of question-wording which is suitable for all purposes. Questions have to be worked out to fit the needs of the particular survey, and, in doing this, the investigator must be quite clear himself about each particular issue, knowing exactly what it is that he wants to discover.

Oppenheim[47] has identified five groups of decisions that have to be made before writing the questions:

I. Those concerning the main and auxiliary methods of data collection; that is, whether by interviews, postal survey, observation, and so on. In particular, it must be decided whether the respondent is to fill in the questionnaire himself, or whether interviewers are to ask the questions and fill in the responses on a schedule, as this is important in the design of the document.

II. Decisions concerning the methods of approach to the respondents (after selection by sampling procedures), including whether or not to indicate sponsorship, the purpose of research, confidentiality, and anonymity.

III. Decisions concerning question-sequences and the ordering of questions within the framework of the questionnaire.

IV. For each variable, the order of questions within each question-sequence and the use of techniques such as *funnelling*, *quintamensional design*, and *factual or attitudinal opening*.

V. Decisions about the use of different types of questions, such as *two-way* and *free-response* questions.

Decisions about these matters should generally be made before the questionnaire is designed.

In all survey work the problem of loss of information arises. The difficulty is not how to avoid loss of information – as this is almost inevitable – but at what point loss of information can no longer be accepted. Information can be lost in interviewing, coding, and statistical analysis; and it must be decided at what stage information loss will cause least bias. It is here that pilot studies can be particularly valuable. They enable the investigator to identify irrelevant questions which can then be omitted from the final questionnaire in the main study.

Good pilot work is essential for the successful use of questionnaires. In the first stages, pilot work is likely to be exploratory, consisting of unstructured interviews and talks. Once a 'feel' for the subject has been gained, the remainder of pilot work can be in organized, experimental forms.

The greatest value of pilot work is, however, in helping to devise the actual wording of questions, in designing the introductory letter, and in finding methods of reducing levels of nonresponse. Questions should preferably be piloted several times until they are as near perfect as possible – although it must be recognized that frequently the opportunity to do this will be limited. It is probably best to divide questions into subsections, each covering a particular variable which is to be explored. Each section can then be tested separately, thus breaking up the pilot work into several small operations.

In principle, a question is satisfactorily worded if the answers tell us what we wanted to know; but there is also the problem of inadvertent bias. A poor question will produce a narrow range of responses; or, perhaps, a very wide range, indicating that it can be interpreted ambiguously. It may be misunderstood by some respondents; it may be too vague, or ask for information that the respondent cannot remember, or does not have; it may be a leading question which biases answers; it may be at the wrong level of intimacy, may be too colloquial or too technical; it may need an introductory or supplementary sentence. Many of these problems can be solved, at least in part, by careful piloting and experience.

A useful tool in pilot work is the *split-ballot technique*. By this method, a question is posed in two different forms to two samples of respondents, matched as closely as possible. Excessive differences in answers can then be assumed to be due to the differences in the wording. An alternative approach is to administer a short questionnaire to a small sample of people. All answers to each question are then compared for differences arising from ambiguities in the questions. Pilot work also enables the researcher to develop 'free-answer' questions into 'multiple-choice' ones for the main studies, by suggesting categories for responses. Pilot samples should use respondents as similar as possible to those who will be approached in the main inquiry. The sequence of questions must be considered before their wording. It must be decided, for example, whether factual or attitudinal questions should be asked first; where the profile questions should be placed; whether any questions should be repeated in different places for corroboration. It is also necessary to consider how the questionnaire can be made interesting and attractive to the respondent and how to establish *rapport* before asking personal, searching questions.

The final form of the questionnaire should be piloted *as a whole*, to test the sequence and the context of each question. The question order should be such as to avoid putting ideas into the respondent's head or suggesting he should have attitudes where he has none.

In addition to the sequence of questions, there is the problem of deciding the type of questions to be asked. Question types fall, broadly, into two groups: *open* and *closed* questions. The closed question is one where the respondent is offered a choice of alternative (predetermined) replies. The open (or free-answer) question is not followed by such a choice, and the answer must be recorded in full.

Open questions are 'easy to ask, difficult to answer, and more difficult to analyse'.[47] A coding frame can be drawn up to classify the answers, but this, and the actual coding, is time-consuming and requires considerable training. Open questions are, therefore, usually limited in number on most questionnaires. The chief advantage of the open question is that replies are spontaneous, free and in the respondent's own words. If the first answer is insufficient or ambiguous the interviewer may be instructed to probe, although this creates the risk of extensive bias. Another risk is that the respondent may simply speak of what is uppermost in his mind: it is difficult to tell from the responses the level at which the respondent was thinking at the time of the interview.

Closed questions can either be *two-way* – that is, having only two possible answers – or *multiple-choice*; and they can be *factual* or *attitudinal*. They are quicker and easier to answer than open questions. They require no writing and quantification is simple. This often means that more can be achieved within a given budget. But there is a loss of spontaneity, and a possible introduction of bias by offering alternatives which the respondent would not have

57

thought of for himself. Closed questions are usually simpler than open questions, and there may be some loss of rapport if the respondent feels that the proffered alternatives do not give proper scope for his ideas. Closed questions should allow for those who appear to feel neither one way nor the other about a given subject; that is, sometimes an 'undecided', 'about the same', or 'both' box should be supplied. It is also sometimes helpful to add a category for qualified 'yes' or 'no' answers. On the other hand a middle category may allow leanings to be obscured, as people not feeling strongly will tend to choose this category; or it can be used to sort out those with strong convictions only.

Closed questions often make use of a check-list from which the respondent selects his answer. This is a help when there are several possible answers and it is unfair, and counter-productive, to ask the respondent to remember the whole of a long and detailed list. Such a check-list is also helpful in establishing the categories and terms in which answers are desired, and eliminating miscellaneous and uncodable answers. It has the disadvantage, however, that it limits the range of possible responses. And, moreover, empirical evidence has shown that respondents tend to choose alternatives near the middle of a list *if the given alternatives are numerical*; with a list of *ideas*, they tend to choose those near the top or the bottom of the list. The latter problem can be solved by randomizing the order in which the ideas are presented thus cancelling out the biases. But, with lists of numbers, the possibilities of bias simply have to be accepted as inevitable. This is because, in general, a random list of numbers tends to confuse respondents. The split-ballot technique may also be useful here; the sample can be divided into two or more equivalent parts and each presented with a different answer sequence, which makes it possible to measure and allow for ordinal bias.

A check-list must be exhaustive if the range of possibilities is to be adequately covered, since few respondents will volunteer ideas not shown on the list, whatever they may think. Careful pilot work is needed to give an adequate range and to ensure that the terms used have roughly similar meanings to all respondents. Similarly, multiple choices must be properly balanced: a list must not include the same idea twice in different form, nor must closely related ideas be offered as a choice against unrelated ones. Mutually exclusive categories must be used.

The two different types of questions – open and closed – have different purposes and thus one kind cannot necessarily be said to be 'better' than another. Open questions will usually be employed only in order to explore the issue in hand. Closed questions, on the other hand, can be used to bring all sides of a question to the respondent's attention.

There are various techniques by which both kinds of questions can be used on a single questionnaire, perhaps the most usual being the *filter* or *funnel* approach. This method is useful for assessing a respondent's knowledge of a subject before asking what he thinks of it and in excluding a respondent from a particular question sequence if those questions are irrelevant to him. A respondent can be asked, for example, whether he ever takes trips into the countryside; if he does, then the interviewer proceeds to ask about the frequency of trips, their location, the activities that are engaged in, and so on. A general question is asked first, followed by more detailed ones as appropriate: 'Have you heard of the Lickey Hills? What recreation activities go on in the Lickey Hills? Do you ever go there for recreation? What do you do there?'; and so on. In general, the principle of filter questions is to proceed from the general to the specific, from open to closed – structured and precoded – questions.

A second technique which utilises all types of questions is Gallup's *Quintamensional Design*.[18] According to Gallup, five questions can be used to cover most of the essential features of an opinion or attitude, as follows:

Question 1: designed to find out whether the respondent is aware of the issue at all, or if he has thought about it; e.g.: 'Will you tell me what planning permission means to you?'
Question 2: designed to get general feelings on the issue, and always an open question: 'What is your general attitude towards planning controls in this country?'
Question 3: designed to get answers on a specific part of the issue – done by precoded questions: 'It has been suggested that Green Belts can no longer serve any useful planning purpose. Do you agree or disagree with this?'
Question 4: designed to find out the reasons for the informant's views: 'Why do you feel this way?'
Question 5: designed to find how strongly they are held 'How: strongly do you feel about this – very strongly, fairly strongly or not at all strongly?'

There is one other group of questions that merits brief mention here: the classification or *profile* questions, which constitute a special group of factual questions. Usually, unless quota sampling is used, such questions are put at the end of the questionnaire, when

59

rapport is fully established and the respondent is convinced of the value of the enquiry. Careful attention must be paid to definitions in these questions (for step-children, part-time occupations, etc.). If necessary, supplementary questions must be asked (particularly about jobs) in order to keep 'unclassifiables' to a minimum. The types of profile data that should be sought, and the ways in which they should be sought will be considered in a later chapter.

Having considered in general the sequence of questions, attention should be directed to the questions themselves and their wording. What questions are to be asked will be determined by the objectives and scope of the study. There are, however, some general rules about question-wording.

All questions present problems which arise because some words have different meanings for different respondents: both piloting and careful definitions and instructions to interviewers must be used to overcome this. Words such as 'should', 'could' and 'might' cannot be treated as synonomous. 'Should' generally poses an issue of opinion, 'could' poses possibility, and 'might' probability; and answers will reflect these differences. Again, questions must not take too much for granted: that people have heard of a subject, that they understand it, that they understand the questions, and so on. It must also not be assumed that people always have information; they may not wish to reveal their ignorance, they may be anxious to please, or they may make guesses. 'Face-saving' alternatives and questions that provide an easy let-out are advantageous. The problem is most acute in postal questionnaires, where 'don't knows' may be disguised by nonresponse or by the respondent discussing the question with someone else before writing his answer.

In wording a question which may be interpreted widely, it is necessary to indicate the scope and the terms of an acceptable answer. A 'how much' question by itself will produce a great range of answers, unless an indication is made that answers should be given in miles, minutes, pounds and so on. Sometimes, it helps to set up frames of reference for respondents, since many frequently find it difficult to differentiate, for instance, between large numbers, or to distinguish between the orders of magnitude of towns and cities. One way of helping them is to give a range of numbers on a card from which to choose; but, also, to provide a general frame of reference within which these numbers can be set. For example, a question referring to a particular district within a city might read: 'Which of these other districts would you say it comes closest to in the amount of open space it has – Elm Park with 220 acres, Oakwell with 50, or Ashton with only 10 acres?' This technique has not been much used yet, and little is known about the way responses are influenced. It is generally true, however, that where people have standards of judgement resulting in stable frames of reference, the same answer is likely to be obtained irrespective of the way questions are asked.[50] On the other hand, where people lack reliable standards of judgement and consistent frames of reference they are highly suggestible to the implications of phrases, statements, innuendoes and symbols of any kind, which may serve as clues to help them make up their minds. In the latter case, are the responses really useful anyway? How is one to decide on the 'correct' wording, and can one attribute anything but spurious stability to the answers?

Opinion questions tend to be more sensitive to changes in wording than factual ones. There are two possible approaches to such questions: either simply to enumerate what proportion of respondents say they subscribe to a given opinion, or to attempt to measure the intensity with which people have feelings about the subject. By varying the strength of the stated alternatives some measure may be gained of the number of people who feel strongly about an issue. A strongly worded set of alternatives helps, for instance to 'filter' only those who feel strongly enough to be able to accept the given wording, and leave out those who are relatively neutral. Successive 'eliminator' questions can sometimes be used here. First the respondent is asked to choose between two simple alternatives, and then follow-up questions are used to break down the opinion he chose, while maintaining the other. If this is used, it should be used for both sides of the original question, since it reduces the original opinion until only the most stout adherents remain.

Frequently, also, a questioner assumes that the negative side of a question is so obvious that it need not be stated. This is mistaken, since, unless all the alternatives are stated in full, undue weight is given to one side of the question. It must never be assumed that an *implied* alternative is sufficient.

By the split-ballot technique, 'tight' and 'loose' questions can be isolated; that is, those which get the same replies whatever may be the order in which the alternatives are stated, and those which get statistically significant differences in replies. In 'loose' questions the last alternative given generally seems to be favoured. Simplicity and brevity seem to be the important elements in 'tight' questions. On the other hand, stability of replies to a pretested question is not necessarily a guarantee of good wording: meaningless questions can produce consistent replies. Furthermore, with some questions a high response should not be expected. A choice should not be forced where the basis for one does not really exist.

In some circumstances, the introduction of deliberate misinformation can be useful. Some investigators are opposed in principle to introducing false information unless absolutely necessary as a control, because they believe that the respondent will feel that he is being 'tested' or 'examined' by a slick interviewer. There are, however, occasions when the technique has potential value:

 I. To test response to premises, summaries and expectations; a respondent may correct an error of fact.
 II. A deliberate use can be made of questions which push respondents towards a particular response. Such questions may identify respondents with extreme opinions because they resist the push.
 III. Misinformation can produce unanticipated responses if the respondent feels he should 'straighten out' the interviewer.
 IV. 'Sleeper' questions, about such things as non-existent local planning issues, can help the researcher to gain an insight into the extent of guesswork in responses.

In these and similar circumstances misinformation can be useful. But the interviewer should always attempt to make the misinformation appear genuine. If the respondent catches hold of the idea that he is being asked 'trick' questions, he is likely to become antagonistic and, sometimes, even aggressive.

Problems always arise in trying to put a question in such a way that it is understandable to all respondents without appearing to talk down to some or to mystify others. Here it can be helpful to reverse the term and the explanation; that is, to give the explanation first and then the term, which sounds less patronizing. For example, it might be better to say 'How do you feel about the amount you have to pay for the provision of local services and amenities – your rates, that is?' than the reverse way round which would 'talk down' to some respondents. A further source of trouble may be the confusing antecedent. Unless there is no chance of it being mistaken, the antecedent should be repeated in follow-up questions, so that when asked 'what are your reasons for this?' the respondent does not reply 'This what?'

Questions which sound alike to the respondent but which cover fine shades of detail for the investigator should generally be well separated. If close together, they may provoke nonresponse, the respondent feeling that he has already just been asked the same point.

If extensive questions are to be asked, it is better to split them up, and ask a number of shorter questions. This helps people to remember items, and improves the estimates given, as well as defining the subject in hand. For example, if questions are to be asked about recreation, it is better to split the subject up into separate questions about organized team games, indoor sports, social activities, and so on. Related to this is a general point about the length of questions. At all times it is best not to overload the respondent too heavily, either by asking him to remember too many alternatives, or by making the whole questionnaire too long.

In general, the main problems of question-wording are associated with the attempt not to ask leading questions; that is, to avoid giving respondents the impression that a certain reply is expected, or that some type of answer is preferable to others. This means that emotive, and especially symbolic, words and phrases – for example, 'Would you agree that there is far too much planning these days?' – should be avoided; no implication should be made that knowledge is preferable to ignorance, and so on. In some cases, sensitive subjects have to be examined, and matters of strong opinion, emotion, or prestige have to be discussed. It may be impossible to ask completely neutral questions, or to avoid some respondents wishing to give a good impression of themselves, their knowledge, jobs, or possessions. In such cases all the investigator can do is to word his question as best he can recognizing that problems of this sort will arise; and then to interpret the responses with appropriate caution. There are, however, a few cases where *deliberately loaded* questions have a purpose. Kinsey[34] used them in order to put the onus of denial of certain sexual practices upon the respondent, and to show the respondent that such practices were common and were acceptable admissions. Other occasions have already been mentioned, such as using loaded questions to draw strongly opinionated respondents into giving their views. But the use of loaded questions requires considerable skill in interviewing and should not be undertaken lightly. There are always problems – for example, loaded questions and words mean different things to different people, and responses may well not be comparable. In general, therefore, loaded questions, in whatever form they appear, are to be avoided. Even without them the interpretation of results is difficult enough. Usually, loading merely brings added complications.

The major criterion for assessing questions is whether they are reliable and valid. *Reliability* refers to consistency – that is, obtaining the same results several times over. *Validity* tells whether the question or item really measures what it is intended to measure.

Reliability in factual questions can be ascertained by internal

63

checks. Some of these have been mentioned; they include asking the same question in different forms, split-ballot tests of forms and wording, and the introduction of non-existent items. Validity can be cross-checked by using independent sources of information. It must be remembered, however, that respondents' memories may be faulty, even on the most commonplace matters, and that their interpretations of events will differ. If two informants give different replies about the same event, there is no obvious way of knowing which, if either, is the more 'correct' or valid.

Reliability and validity in attitudinal questions is much more difficult to ascertain. Reliability cannot be checked by asking the question in another form since the response to attitudinal questions is much more sensitive to changes in wording, context and emphasis. The use of sets of questions around a subject may help, as the underlying attitude will tend to be the same for all questions, and, thus, relatively stable elements may be maximized while reducing instability due to particular, momentary items. The difficulty in assessing validity in attitude questions is the lack of criteria. What is needed is groups of people with known attitude characteristics, to see if questions can discriminate between them or not. This is very nearly impossible as behaviour cannot necessarily be predicted from attitudes, or vice versa.

Three sets of people have to use questionnaires and schedules; interviewers (or respondents themselves), coders and punchers. Postal questionnaires obviously have to be constructed differently from schedules for interviewers. Layout and design for both is mostly a matter of common sense; but, for schedules in particular, every consideration should be given to the convenience of interviewers, who will have to handle and use each schedule for a longer period than the coders and punchers. Attention, therefore, has to be paid to ease of marking and coding the responses; the layout of instructions and briefing for the interviewers, both generally and for each individual question. In addition, the work of the coders and punchers has to be considered. Although the time they spend on each schedule is a fraction of that spent by the interviewer, if a puncher, for instance, has to waste several minutes turning over schedules in order to work on them, this does represent a relatively serious loss of time. The convenience of all users should be balanced as far as possible.

Care should also be taken to see that all interviewers interpret the questions in the same way. This is partly done by the verbal briefing of interviewers. It can be assisted by the wording and design of the schedule itself, which can include definitions of terms used, limitations and definitions of subjects discussed, and so on. For postal questionnaires it is important that all respondents should interpret questions in the same way. Postal questionnaires have also to be designed in order to maintain rapport with the respondent in the absence of an interviewer. Attractiveness of layout, wording of introductory letters and phrases, ease of finding one's way about the document, are all elements in maintaining this rapport.

In conclusion, it is worthwhile to repeat and emphasize a few main points about questionnaire design. The main requirement is that the intentions and needs of the investigator be clearly defined. These intentions are the yardstick by which the effectiveness of question wording can be judged. Pretesting and wording must be seen as two aspects of the same problem. The only way to reach any firm conclusions about a particular choice of wording is to test it on a representative sample of respondents. Wording itself is essentially a question of eliminating undesirable biases from the form of the question. The aim is to find out what people really do or think. Finally, design should be such as to give every assistance to those who will use the questionnaires or schedules – principally, respondents or interviewers, but also punchers and coders.

Chapter 5

PROFILE DATA

All social surveys, whether self-administered or by interview, collect information about the characteristics of the survey population. This information, or profile data, covers such items as age, sex, marital status, occupation, income and social class. One major purpose in collecting such data is to provide comparability between the findings of different surveys, and between surveys and published statistics such as the Census. This general purpose has not, however, been much achieved in practice. A major difficulty with a large proportion of recent social research has been a lack of comparability between studies. A recent analysis by Burton and Noad[6] of the profile data collected for six major surveys of leisure patterns showed that all six were comparable on only four characteristics out of a possible total of about forty items. Clearly, it is not necessary that all surveys should seek to obtain data about all possible profile characteristics; but, equally clearly, there ought to be some general agreement about which characteristics should be included, and about the ways in which information concerning these characteristics should be sought. The purpose of this chapter is to identify the major profile characteristics about which data are usually sought, and to suggest guidelines as to the ways in which these data should be collected. The major characteristics are sevenfold:

    I. age;
    II. sex;
    III. marital status;
    IV. family and household;
    V. education;
    VI. income;
    VII. occupation (or social class).

In most surveys the data collected about these seven characteristics are organized into groups; for example, into five marital status groups, or five social classes, or ten income groups. There are, therefore, two distinct but interrelated problems in the collection of profile data; first, what data should be collected; and second in what ways should these data be grouped. Both of these problems will be discussed in relation to the seven characteristics listed above.

Age

There has been almost universal agreement among research investigators that information about age should be collected in all social surveys. Where disagreement has arisen, however, is in the ways in which this information should be collected and the groups into which data about age should be placed for analysis. There are two basic criteria upon which groupings should be determined: first, the subject matter of the survey; and second, comparability with other sources of data. In considering the subject matter of the survey, we should aim at those age groupings which appear to be most appropriate to the content of the survey. Thus, for example, a survey of participation in sport and physical recreation may require very different age groupings from one which is concerned with, say, housing preferences. On the other hand, comparability with other sources demands that we use standardized age-groupings so that, again for example, the relationship between age and leisure patterns may be compared with the relationship between age and, say, health needs. It is, of course, quite possible – indeed, quite likely – that the requirements for satisfying these two separate criteria may conflict: that is, that generally accepted, standardized age groupings may be inappropriate for a survey about a very specific subject or problem.

The subject matter of a survey may require the use of age groupings which are determined by behavioural changes of one kind or another. These behavioural changes may be determined, in turn, by such factors as social custom, administrative procedures, legal requirements, institutional practice, and so on. For example, the law dictates that certain legal responsibilities cannot be assumed until a person reaches the age of majority. Again, most universities refuse admission to undergraduate courses unless the applicant is, at least, 17 years of age. Yet again, many motor insurance companies place higher premiums and greater restrictions upon drivers below the age of 25 years than on those above this age. These and similar behavioural restrictions should be fully considered when preparing the groupings to be used in any particular survey.

The second basic criterion for the determination of age groupings is that they should be comparable with those used in other sources of data. The most important of these sources is, of course, the Census. This gives a breakdown for the national population and for the

67

populations of all local authority areas, by single years of age and separate sexes from 0-20 years, and by sex and marital status in 5-year groupings from 0-4 onwards for all ages. For separate wards, it gives only the total population divided by sex. In addition, the Registrar-General's annual estimates of population are given for each local authority area; and, by sex and 5-year age groupings, for all the standard regions and conurbations.

Many age groupings in current use appear, however, to have no systematic foundation: 'the commonly accepted groupings are in units of 5 years for no better reason than the number of fingers or toes on a hand or foot'.[8] In particular, there are many groupings which assume that there are no significant differences among persons over the age of 65 years. There may, of course, be the elements of a systematic foundation for specific groups in surveys which are concerned with particular topics: in planning surveys, for example, the age-group 20-40 years could be selected as encompassing virtually all potential child-bearing adults.

Broadly, though, it seems true to say that there is little evidence presently available by which to devise meaningful age groupings for use in social surveys. The general evidence that significant changes take place in social patterns between the ages of about 15 and 24 years further complicates the problem, since this time span covers, at most, only two of the age groupings that are in normal use at the present time for almost all social research. If the general evidence about the relationship between social patterns and age during these years is true, it would seem that, perhaps, four or five age groupings should be distinguished during this short time span. These difficulties all suggest that information about age should, at least, be collected in a form which does not presuppose certain age groupings; that is, that information should be collected about the actual age of respondents rather than by placing them within predetermined age groupings. This, however, leads to another, more practical, difficulty. It has often been argued that to ask a respondent his age, or to place himself within a relatively narrow age grouping, can lead to deliberate falsification on his part. In general, there are advantages and disadvantages in overstating or understating age which change, several times, during life. One known tendency is to give one's age *next birthday* when between the ages of 15 and 20 years, and age *last birthday* between the ages of 39 and 65 years.

All of these difficulties suggest that, in planning studies, we should take nothing for granted. This, in turn, leads us to suggest that we should seek the respondent's actual age at the time of the survey, rather than try to place him within predetermined age groups. In order to avoid the incentive to falsification, however, we should ask for date of birth, rather than age last birthday. We can then reduce this to an age and, later, on the basis of the findings of the survey, build up meaningful age groups.

Sex

Physiologically, there are only two sexes – male and female. Chapman[8] has observed that, sociologically, there may be finer distinctions, particularly in relation to behavioural deviance from what is considered socially to be the norm for each of the sexes; but, for survey purposes, the physiological division into male and female is sufficient!

Marital Status

In most social research, marital status is usually considered in 'administrative' terms. This implies a fivefold breakdown: single, married, separated, divorced and widowed. Chapman[8] has argued, however, that, for research purposes, this breakdown needs extending. He argues that the administrative definitions imply, firstly, that all marriages are socio-biological in character and, secondly, that all unmarried persons do not have a socio-biological relationship. He suggests, therefore, that 'account must be taken of unions which are biological but not hallowed or registered', and that a sixth category – cohabiting – should be introduced.

The implicit assumption behind Chapman's suggestion is that cohabitation would show distinct characteristics from the other five categories of marital status. This may be so for certain legal-administrative characteristics, but it is unlikely that there are strong distinguishing behavioural patterns. It is highly probable that, behaviourally, cohabiting couples are akin to married couples; and as long as such persons describe themselves as 'married', the distinction of a category 'cohabiting' is not necessary.

There is also a practical point here. Cohabitation still carries an element of social stigma and is, therefore, likely to be a highly sensitive subject upon which to question respondents. It is not impossible for a skilled interviewer to obtain such information without antagonizing the respondent; but it can be argued that the risk of creating antagonism should be taken only if the resulting data are essential to the success of the study. It does not appear that the distinction of a category 'cohabiting' would add greatly to the significance of most social studies; and it is suggested, therefore, that the commonly accepted fivefold classification given above is fully adequate for the great majority of research studies.

69

Family and Household

It is clear that a distinction should be made between 'the family' and 'the household' in social research. Conceptually, the distinction is not difficult to make. The household is based upon communal living – that is, the group of people who share a common dwelling, who normally eat together and have common housekeeping. In practice, there is some variation in definition between authors; but, by and large, the Census definition[51] is the most accepted. This has the great advantage that it allows for comparability of data between the wealth of information obtained through the Census, and other studies. The definition is:

I. 'Any group of persons, whether related or not, who live together and benefit from a common housekeeping; or any person living alone who is responsible for providing his or her own meals.

II. A person living, but not taking meals, with a private household was treated as a separate household, but if that person had at least one meal a day with the household he was regarded as part of that household. Breakfast counted as a meal for this purpose.

III. By convention a household had to have at least one room. Two or more persons living in one room were regarded as one household regardless of whether or not they had their meals together or shared a common housekeeping.

IV. A person or group of people living in a non-private establishment were treated as a separate household if they were either –

(*a*) a family which does not normally depend on the institution for the provision of meals, or

(*b*) a person or group for whom the institution does not provide any daily meals.'

Thus, the household is defined on the basis of communal living.

The family on the other hand, is defined on the basis of kinship. The *immediate* family usually consists of a married couple and their children; the *extended* family is composed of a kinship grouping of persons related by blood, marriage or adoption, which is wider than the immediate family. In general, the members of an immediate family occupy the same dwelling and form a single household. (This, incidentally, is one of the primary causes of the prevalent confusion between 'the family' and 'the household' in much social research.)

70

It is by no means established that either the family or the household should be the basic unit for consideration in relation to social studies. Most recent social surveys have obtained information about the structure of the household rather than the family. This has the added advantage that if the information is collected by asking the

Table 1. *Standardized Household Composition Chart*

NO.	RELATIONSHIP TO H.O.H.	SEX		DATE OF BIRTH	MARITAL STATUS				
		M	F		M	W	D	SEP	S
1	H.O.H.								
2									
3									
4									
5									
6									
7									
8									
9									
10									

respondent to list all members of the household according to their relationship to the head of household, the family groupings can be constructed upon the basis of the same data. It is, therefore, argued that this is the most practical approach to take.

The information collected in respect of the four profile characteristics so far discussed – age, sex, marital status and family or household composition – can be collected in the form of a *standardized household composition chart*, an example of which is given in Table 1.

Education
It is by no means universal practice to use education as a key variable in social research. This arises, in part, from a belief that educational

71

standard is reflected, along with other key variables, in social class (or status) groupings, which are much more usually employed in survey work; and, in part, because unlike most other key variables, educational groupings are subject to fairly rapid definitional and compositional change and, consequently have very limited value for intergeneration comparisons. The first of these objections to the use of educational groupings may be considered as conceptual; the second as functional.

Chapman[8] has suggested that income and education, in its broad sense, control all the major variables which determine social groupings: 'These fundamental influences have markedly different characteristics. High income can purchase many of the indices of status, but not all; those that remain derive from education, which is almost always conditioned by the income and education of parents.' This view, has, however, been challenged by a number of writers. Couter and Downham[13] noted that people with a similar education, *as defined by terminal education age*, were much more likely to resemble one another in their sports interests and habits, irrespective of 'class' than to resemble people of their own 'class' with a different education. Abrams[1] has suggested that the relationship is more complex than is popularly assumed. In his study of reading habits in Britain, he compared two very broad age groups – 20–44 years and 45 years and over – in terms of three terminal education ages. He found that differences in social class (as defined by the Registrar-General) for persons over 45 years of age are rarely the outcome of differences in educational background, whereas 'among adults under 45 years of age, differences in education play a larger part in differentiating the social classes'.

Virtually all writers accept the view that educational standards play a part in determining social groupings. But then, so do income and occupation; and, sometimes, family background and racial characteristics. The argument is not, therefore, about whether education is related to social status, but whether its influence upon patterns of behaviour and life styles is independent of the influence of social status, or whether it operates entirely through the latter. In other words, are differences in educational standard adequately reflected in differences in social status? The answer depends in part, of course, upon what characteristic, or variable, is taken as the measure of social status. It also depends in part upon the subject matter which the researcher is investigating. Educational standard may have no effects, independent of social status, upon shopping preferences as between say, the family grocer and the supermarket. It may, on the other hand, have a significant independent influence upon leisure habits, irrespective of the influence of social status. From the planning point of view the assumption is that the level of educational attainment has some bearing on, for example, rates of participation in recreation pursuits, because of an association with aspects of educational curricula and exposure to certain types of novel experience, such as tuition in the arts, golf or horse riding: 'In recent years, thanks largely to the patronage and participation of younger people who have enjoyed wider and more secondary education than their parents, there has been a marked upsurge in interest in all the Arts, at both the professional and amateur level, both in audiences and performers.'[74]

It is significant that none of the five methods most commonly used for defining social groupings* is based directly upon educational standard. The most generally accepted – though not necessarily the most correct or accurate – criterion of social grouping is occupation. And, although this may reflect, in part, educational standard, there are clear indications that the relationship between the two is not entirely consistent. Thus, for example, Weinberg[71] has examined the Tables of Education from the ten per cent sample of the *Census, 1961*, with the Tables of Socio-Economic Groups from the same source. He found a number of unexpected relationships; *Socio-Economic Group 2* – male employers and managers in industry and commerce in small establishments – included 54 per cent who had left school before the age of 15 years, whereas, for *Group 6* – junior non-manual workers – the proportion was only 45 per cent. A similar margin of difference was noted between *Group 8* – foremen and supervisors, manual – and *Group 11* – unskilled manual workers; the latter group included 75 per cent who had left school before the age of 15 years, while the proportion for the former group was 83 per cent. Weinberg accepts that there are often plausible explanations for differences of this kind; but he argues that, nonetheless, 'there is little evidence here that social class (as measured by occupation) is directly related to educational background, at least as measured by school leaving age'.

The evidence is, then, that social status is not solely a function of educational standard; or even, that social status or grouping, while being a function of several variables, reflects educational standards consistently. This suggests that data concerning educational standard should be sought directly from respondents rather than assumed in data on social grouping. This proposal is further reinforced if there is reason to suppose that educational standard will have a direct and independent effect upon the particular behaviour

* See below, p. 81.

73

that the researcher is investigating – in addition to any effects which might operate through social grouping (or, indeed, through any other characteristic such as income). There is still the problem, however, of which of various methods of measuring educational standard should be employed.

Weinberg's analysis of 47 community research studies indicated, firstly, that only a limited number of investigators collect data about educational achievement and, secondly, that there are several very distinct methods of measuring this achievement.[71] Of the 47 studies which he examined, 26 made no reference of any kind to education. A further 7 published questionnaires which included questions on education but presented no data at all concerning this. One study made a general (descriptive) reference to it, while the remaining 13 related it to at least a few other variables.

The most common method used for classifying education among Weinberg's studies was participation in further education of some kind after leaving school. There was, however, a notable lack of comparability; only 6 studies requested data about further education in all its forms and only 5 distinguished between full-time and part-time study.

The second most commonly used method was to relate educational standard to terminal education age. By this was usually meant terminal *school* age. This, of course, provides a diverse mixture of different kinds of leavers at age 15 years – for example, secondary modern pupils, early leavers from grammar schools, leavers from all streams of comprehensive schools, leavers with some GCE Ordinary Level qualifications and others without, and so on. Thus, terminal school age, by itself, succeeds merely in distinguishing the number of years of formal schooling above the statutory minimum. It tells us nothing about the type of schooling.

Some studies did, however, record the type of school attended – a few of which did this in addition to collecting data about terminal education age. There are particular difficulties of comparability here; firstly, as between generations, because of substantial changes in the education system itself – for example, comprehensive schools were virtually non-existent in the 1950s; secondly, as between schools of the same category in different parts of the country, since standards often vary from one part of the country to another; indeed, some schools with the same title have very different functions in different parts of the country – for example, in some areas comprehensive schools are replacing grammar schools, in others they are simply an alternative.

Weinberg's analysis indicates, then, that the two criteria which are

74

most commonly used by social investigators are (I) participation in further education and (II) terminal education (that is, school) age. In addition, some investigators recorded the type of school attended; and in three cases, the educational background of the parents was requested. The problem is that we do not know which of these different measures is the most satisfactory. Moreover, there is not simply the problem of choice between three or four indicators of educational standard; it is also clear that there are a number of different ways in which each indicator can be measured.

The first major indicator is the type of school attended; and there would appear to be three ways in which this can be distinguished: firstly, by the type of operator – for example, local education authority, Roman Catholic Church, and so on; secondly, by the type of school itself – for example, grammar and comprehensive; and thirdly, by whether the school is for single-sex pupils or is coeducational.

The *type of operator* is particularly important in certain areas of the country. In Liverpool, for example, the denominational schools (that is, mainly Catholic) play an important part in maintaining the fairly rigid identity of local communities. It is, however, only the Catholics who are actively expanding the numbers of their secondary schools, so that it is perhaps only they who should be considered as a distinct type of religious school. A second type of operator to be considered is the private operator. The third, of course, is the local education authority. The suggested breakdown for type of operator is, therefore, fourfold: (I) Local Education Authority; (II) Roman Catholic; (III) Other Religious; and (IV) Private (and other).

The *type of school attended* would be based upon administrative definitions. This does assume that a comprehensive school in one local authority area is comparable with another comprehensive school elsewhere. Unfortunately, however, there does not seem to be any alternative to this, and anyway, although not wholly true throughout the country, the administrative classifications are generally comparable over most of the country. The suggested breakdown for type of school is therefore fivefold: (I) Elementary (that is, pre-1947) and Secondary Modern; (II) Grammar and Technical; (III) Comprehensive and Bi-lateral; (IV) Public and Private; and (V) Other (for example, Special Schools).

When the distinction between single sex schools and coeducational schools is built into these groupings, the final table for type of school attended would appear as shown in Table 2.

Terminal school age may be expressed either as an actual age or as a period of education beyond the statutory school leaving age. If actual age is asked, it will be possible to calculate the period of extended education, but only in approximate terms; that is, if a person terminates schooling at age sixteen, then his period of education beyond the statutory school leaving age (as it stands at present, that is fifteen years of age) could range from two school terms to five school terms. If, therefore, more detailed information is required, it will be necessary to inquire about the *period* of schooling beyond the statutory school leaving age. There is another problem here: the statutory school leaving age has changed once since the Second World War and is likely to do so again soon. This suggests that, for inter-generation comparisons, it is necessary also to enquire about actual leaving age. Weinberg's suggested classification, with modifications, is shown in Table 3.

Further Education may be considered in at least four ways; by type of institution attended; by whether attendance is full-time or part-time; by the duration of attendance; and by the qualifications

Table 2. *Suggested Classification for Type of School Attended*

	LEA		RC		Other Religious		Private and Other	
	ss.	co-ed.	ss.	co-ed.	ss.	co-ed.	ss.	co-ed.
Elementary/Modern	1	2	3	4	5	6	–	–
Grammar/Technical	1	2	3	4	5	6	–	–
Comprehensive/	1	2	3	4	5	6	–	–
Bilateral	1	2	3	4	5	6	–	–
Public/Private	–	–	3	4	5	6	7	8
Other	1	2	3	4	5	6	7	8

Source: Modified from A. Weinberg,[71] *op. cit.*

Table 3. *Suggested Classification for Terminal School Age*

	Statutory Leaving Age		
	14 years (or less)	15 years	16 years
Premature Leaver before age of	1	2	3
Left at earliest opportunity after reaching leaving age	1	2	3
Stayed on One term	1	2	3
Two terms	1	2	3
One year	1	2	3
1–2 years	1	2	3
2–3 years	1	2	3
3+ years	1	2	3

Source: Modified from A. Weinberg,[71] *op. cit.*

obtained. The important measures are, we suggest, the second and third of these. Whether the course is part-time or full-time and the number of years over which it takes place probably have a significant effect upon a wide range of social issues and attitudes. The qualification being sought (that is, degree, certificate of education, higher national diploma, and so on) and the type of institution (university, college of education, technical college, etc.) are also significant but probably less so. In particular, it is worth distinguishing between further and higher education; the former being defined as a course leading to a qualification which is of an *equivalent* standard to qualifications that can be obtained at school – that is, GCE Ordinary and Advanced Levels. Higher Education involves courses leading to qualifications which are of a *higher* standard than those which could be obtained at school. There are, obviously, some difficulties arising from the standard of qualifications for subject areas which cannot be studied at all at schools – for example, secretarial qualifications. The report of the Committee on Higher Education,[12] in Appendix 1, Annex D, gives detailed information concerning these and other professional qualifications. The suggested grouping for the data is shown in Table 4.

Table 4. *Suggested Classification of Higher and Further Education Courses*

Further Education Courses

	Duration of Study				
	1 year	*2 year*	*3 years*	*4 years*	*5 years*
Full-time for basic qualification	1	2	3	4	5
Part-time for basic qualification	1	2	3	4	5
Full-time for higher qualification	1	2	3	4	5
Part-time for higher qualification	1	2	3	4	5

Some studies have sought to obtain data about parents' education since it is generally agreed that this will often influence children's attitudes and behaviour. There are particular difficulties in doing this, however, when conducting a survey of all age-groups within the general population, since, several investigators have found a large proportion of respondents do not have the requisite information. This makes inter-generation comparisons difficult. It may be worth attempting to secure this information, for both parents, where the subject of the survey makes it likely that it could be very significant in the findings. But, in view of difficulties of comparability and practicability, it cannot be recommended for general use in all surveys.

Income

The collection of income data in social surveys presents two broad categories of problems – those of definition and those of method. Problems in the former group include such questions as how income is to be measured and whose income is to be considered. Those in the latter group are concerned with such questions as how to obtain accurate and sufficient data about income. In addition, there is a third – part definitional, part methodological – problem; namely, whether or not it is possible, for survey design and analysis purposes, to seek data about incomes in the form of defined income groups.

There is, however, a fundamental prior consideration to these – though one which is, surprisingly, often ignored. A first question should be whether or not it is *necessary* to obtain income data; or, alternatively, whether income data alone will be *sufficient* for the purposes of the survey. This latter question is particularly important since it is generally assumed, in practice, that data about some aspect of income is always necessary. It is much more questionable, however, whether income data alone will be adequate. It may be necessary, for example, to seek data about patterns of expenditure rather than levels of income.

Gittus[22] has identified four main reasons for including questions about income in a survey. They are, firstly, as part of the background information with little or no theoretical application in descriptive surveys; secondly, in industrial studies, as a means of relating satisfaction with earnings to such characteristics as the level of morale; thirdly, in 'social accounting', as a precise variable to be related to certain socio-economic factors, such as expenditure patterns; and fourthly, more analytically, as one of the variables to be considered in a discussion of social stratification. All of these are reasons why it has been thought necessary to collect data about income, but only in the second case (industrial studies) is the collection of income data alone likely to be sufficient. The collection of income data in most social surveys would be necessary for Gittus's first and third reasons – as part of the background information about respondents in descriptive surveys and as a basis for some element of 'social accounting'.

A further important consideration is comparability with other studies. There are a number of regular and detailed studies of income and expenditure made in Britain – in particular, the annual Family Expenditure Survey of the Department of Employment and Productivity (formerly the Ministry of Labour) – which could provide a useful source of comparative data so long as definitions and meanings are comparable.

We turn now to definitional problems, of which there are basically three: first, the unit – for example, individual or household – for whom income data will be sought; second, the period of time to which this income data should refer; and third, the sources of income that are to be included.

The choice of income unit is not straightforward. If comparability with official (and other) available data were the main criterion, then the household would be the most frequent choice. This is the basic unit of analysis in ths Family Expenditure Surveys.[39] On the other hand, the Ministry of Social Security's *Circumstances of Families*,[41] is primarily concerned with the combined resources of the husband and wife. Both ministries, however, collected their data for each adult spender separately. This enabled them then to build up the data into units based upon the household on the one hand, and the husband and wife on the other. Thus, it was only the unit of analysis which was dissimilar.

The household unit may, however, be unsatisfactory for particular surveys. There is, for example, substantial empirical evidence which suggests that recreation habits and patterns of activity are only occasionally based upon household groups. A more significant variable is age, which causes habits and activities to cut across households. This suggests that the unit for recreation surveys should be the individual; units can then later be amalgamated into such groups as families and households.

There is still, of course, the problem of how many individuals' incomes should be collected. Generally, data would be collected only for individuals who are gainfully employed, irrespective of age. The Family Expenditure Surveys defined a person who is gainfully employed as one who is employed for more than ten hours a week, or is an employer or self-employed. Broadly, this means that data are collected only for incomes brought into the household or family from outside sources – although it does not include all incomes of this nature. It does, however, exclude payments from one member of the household or family to another – for example, housekeeping money, dress allowance and pocket money.

This leads to the second main problem – the sources of income that are to be included. Gittus,[22] in discussing industrial studies, suggests that, in principle, a distinction should be made between (I) the basic wage, (II) the 'take home' pay (or net earnings), and (III) the 'normal' (or usual) earnings: 'Data should be obtained on variations in (I) . . . and also on variations in the difference between (I) and (II), or (I) and (III), and in the factors . . . that account for this difference.' This refers, however, only to incomes from employment. In wider studies, account would need to be taken of such sources as profits, dividends and rents. Writing of Banbury, for instance, Stacey[59] distinguishes between (I) profits and fees, (II) salaries and (III) wages; the distinction between (II) and (III) being, presumably, that the former are paid monthly and the latter weekly.

The Family Expenditure Surveys were more detailed. They distinguished eight main sources: wages and salaries of employees; self-employment income; income from investments; income from non-State pensions and annuities; State retirement, old age and widows' pensions; other State benefits; income from sub-letting and/or owner-occupation; and income from other sources. Each of these major sources is divided into component sources as shown in Table 5. Specifically excluded were money received by one member of the household from another; withdrawals of savings; maturing insurance policies; proceeds of the sale of houses, cars, etc.; windfalls, such as legacies; winnings from gambling; and the value of educational grants and scholarships and of concessionary goods received free or at reduced prices from employer or state.

The third major definitional problem concerns the period of time to which income data should refer – usually, week, month or year. The unit of analysis in most previous studies has been the year. In many studies, it has also been the unit of collection, respondents being asked to indicate approximately, to which of several given (per annum) income groups they belong. In the Family Expenditure Surveys, however, there has been a different approach. The data are collected for the periods to which they commonly referred. Thus, wages are recorded in amounts per week, salaries per month, dividends biannually, and so on. These are then converted to a common period (per week) in the analysis. This has the advantage that respondents are being asked to supply data for periods for which they commonly 'budget' (in the broad sense of the word). Difficulties arise, however, in eliciting data from weekly-paid workers about overtime and bonus payments. For simplicity the study adopted the category of 'normal take home pay'.

Methodological problems are concerned with ways and means of obtaining accurate and sufficient data about income. If, for example, the respondent in a survey is the head of the household, it is usually not difficult to obtain income data pertaining to the whole family from him. But, if any other member of the household should be the respondent, then experience has shown that they often do not have detailed information about the incomes of other members of the household. This type of methodological problem is fairly straightforward and can be largely solved by ensuring that, where income

80

data are required from more than one member of the household or family, the respondent is the head of the household.

A more difficult problem often arises with self-employed persons and with those receiving incomes from dividends and similar sources. Some self-employed persons are often just wholly ignorant of their financial affairs, leaving these in the hands of an accountant or bank manager. Many more persons, who receive biannual or quarterly dividends and interest, are often unsure whether this income is gross or net of income tax. Utting[67] found this to be especially true of retired people living on pensions and small dividends.

A third methodological problem is, of course, simply to obtain any kind of income data at all from respondents. Many people are reluctant to supply income data to (unknown) interviewers. Refusal rates for questions about income are generally higher than for questions on any other topic in household surveys. This type of problem is not very easily solved. In the long run, it depends, at least in part, on the training and experience of interviewers. With a well-trained and experienced interviewer, it is often best to frame income questions very generally and then to leave the interviewer free to probe in varying ways for different respondents. The usual approach to this problem has been to reduce the amount of 'personal' information sought by asking respondents to indicate particular income groups within which their own incomes fall rather than to ask for actual incomes. This, of course, requires that the research investigator draws up a list of predetermined income groups. This, in turn, requires that the investigator has prior knowledge of critical 'margins' between levels of income which are significant to the subject matter of the survey. Yet, there is often no information available to indicate what are these critical margins. The major reason for choosing particular groupings is, usually, therefore to obtain comparability with other studies. In the absence of any better standard, this is clearly the most useful approach. The groups most commonly used at present are shown in Table 6.

Social Groupings and Occupation

There are several alternative definitions of social class and socio-economic grouping currently in use for survey research in this country, none of which has received clear acceptance from the majority of research workers. As a result, considerable confusion and a lack of comparability between surveys have arisen. Even when several investigators have ostensibly used the same grouping in their surveys, there has been a tendency for definitions to differ quite significantly in practice.

F

Table 5. *Department of Employment and Productivity, Family Expenditure Survey - Income Groups*

(1)	(2)	(3)
Item Group	Main Sources of Income	Components of Main Sources of Income
(a)	Wages and salaries of employees working more than 10 hours a week	Normal 'take-home' pay Employees' income tax deductions Employees' National Insurance contributions Employees' Graduated National Insurance Scheme deductions Other deductions
(b)	Self-employment income	Income from business or profession
(c)	Income from investments	Interest, after tax, on building society shares and deposits Interest on co-operative society shares and deposits including dividends on purchases Interest on bank deposits and savings accounts, including Post Office Savings Bank Interest on Defence Bonds, National Development Bonds and War Loans Interest and dividends, after tax, from stocks, shares, bonds, debentures and any other securities. Income from trust or covenant. Rent or income from property (not own residence) after deducting expenses allowed for income tax Other unearned income
(d)	Income from non-State pensions and annuities	Pensions from central or local government services or from the armed forces Other pensions Income tax deducted from pensions Annuities
(e)	State retirement, old age and widows' pensions	Retirement pensions Old age or widows' pensions

(1)	(2)	(3)
Item Group	Main Sources of Income	Components of Main Sources of Income
(f)	Other state benefits	Family allowance Disablement pension War disability pensions or allowance Unemployment benefit Sickness benefit Industrial Injury Benefit Supplementary pension or supplementary allowance Any other benefits
(g)	Income from sub-letting and/or owner-occupation	For rented (or rent free) dwelling, any excess of rent received over rent, rates and water charges, etc., paid out For owner-occupied dwelling, either rateable value of dwelling occupied or the excess of rent received over the sum of ground rent, rates and water charges and insurance on dwelling whichever is the greater
(h)	Income from other sources	Wage or salary of person working 10 hours a week or less Married woman's allowance from absent husband Allowances from members of the armed forces or merchant navy (excluding husband) who are not members of household Money scholarship disbursed by member of household completing a personal schedule Alimony separation allowances or any other money from friends or relatives outside the household Trade Union benefits Friendly-Society benefits Other non-State benefits Other income from other sources Money scholarship disbursed by member of household not completing a personal schedule Other income of member of household not completing a personal schedule

Table 6. *Commonly Accepted Income Groupings, 1969*

Up to £5 per week (= £260 per annum)
Over £5 up to £7 10s p.w. (= £261–£390 p.a.)
Over £7 10s up to £10 p.w. (= £391–£520 p.a.)
Over £10 up to £12 10s p.w. (= £521–£650 p.a.)
Over £12 10s up to £15 p.w. (= £651–£780 p.a.)
Over £15 up to £20 p.w. (= £781–£1,040 p.a.)
Over £20 up to £25 p.w. (= £1,041–£1,300 p.a.)
Over £25 up to £30 p.w. (= £1,031–£1,560 p.a.)
Over £30 up to £40 p.w. (= £1,561–£2,080 p.a.)
Over £40 (= £2,080 p.a.)

Social groupings have, broadly, two main functions: on the one hand, they provide a basis for quota sampling, which influences and restricts the selection of respondents for particular surveys; on the other hand, they provide a means of tabulating and presenting data relating to the characteristics of respondents, who have been selected for survey by other, possibly random sampling methods. It has been argued that the second of these functions is the most important; and that their use as a basis for drawing quota samples is merely a subsidiary function that social groupings can perform. This point has been strongly debated, but one thing seems quite clear. This is that the sampling function does not necessarily require the *same* social groupings as the analytical-interpretive function. The problem of defining social groupings is clearly bound up very closely with the problem of choice between alternative sampling methods, but the essence of the relationship lies in the relative adequacy of quota sampling as against other sampling methods, not in the validity of the particular social groupings that are used to determine quota samples.

The objective of a social grouping in survey research should be 'to classify people into groups in such a way that information obtained in one context (say, a readership survey) could be attributed to the members of these groups in another context, say, a market study'.[38] In other words, it should be a social grouping or classification which will serve more than one purpose or survey.

A subsidiary objective is, often, to provide a 'reasonable' indicator of a number of different characteristics of people being surveyed which, taken separately, would require the collection of a large volume of data from respondents, some of which might be difficult to obtain anyway – for example, income. In particular, a social grouping should seek to provide an indicator of variations in both economic and social status, and not just the one or the other.

84

One of the major difficulties that has arisen in relation to the definition of social groupings stems from a widespread confusion over the meanings of the terms 'social class' and 'social status'. On some occasions the terms are used interchangeably; on others, they have had very different meanings. The result has been that many studies have turned out to be both non-comparable and, often, simply confusing.

The only school of thought which has made a clear distinction between social class and social status is the Marx-Weber School. Here, the class situation refers essentially to the extent of control over goods and the possibility of their exploitation: 'Classes are stratified according to their relations to the production and acquisition of goods; whereas status groups are stratified according to principles of their consumption of goods as represented by special styles of life.'[73]

This type of differentiation does give some help in clarifying the distinction between 'class' and 'status' (or 'social group'). Class has a clear relationship to the political and economic structure of the country. It is possible, therefore, to restrict this discussion to status and social grouping in terms of particular variables, such as income, education and spending (that is, consumption) patterns, while keeping clear of the controversial field of politics into which a discussion of 'class' is likely to lead.

One of the important results of this separation of concepts is that a distinction can then be made between what people think about social groupings and their real socio-economic situations; and, perhaps from this, an idea can be obtained of the directions in which changes are taking place. There are two important considerations here. Firstly, in a general sense, and despite obvious limitations, income levels are related to the economic priorities of a country. For instance, the fact that scientists and technologists are very much better paid in relation to other professions in the United States than in Britain does tell us something, however crude, about the importance which we must assign to science and technology. Secondly, status tends to follow, rather than precede, changes in economic position. That is to say, it usually takes longer for a group which is rising in economic importance to gain social acceptance and status than to gain financial reward; similarly, a declining group usually loses income faster than it loses social status. Clergymen are a case in point; they tend to have considerably higher social status than one would normally associate with persons at their income levels. (From these two points, conclusions might be drawn about the changing importance of clergymen as opposed to, say, skilled technicians.)

85

Another of the difficulties with many current definitions of social grouping is that they divide society only into horizontal groups, usually based upon a hierarchy related to income, occupation, education, and so on, or some combination of such characteristics. But society is also characterized by what the anthropologist calls 'cross-cutting ties'. That is to say, society does not fall into simple groups, but into groups which cut across each other. The ties thus created prevent society from falling apart into opposed groups, because people who are a part of one group in one context, may be combined with enemies or opposites of that group in another. For example, men who oppose each other on the football pitch may stand together in opposition to an employer.

We turn now from concepts to methods: that is, from a discussion of the nature and functions of social groupings to a discussion of the various ways in which such groupings are constructed. It is here that the use of 'occupation' in the title for this section is pertinent. In general, occupation is the most commonly used indicator of social grouping. There has been some controversy as to whether or not this is a legitimate basis. The major arguments both for and against this will be discussed in the context of the various different methods of grouping. These methods are, basically, fivefold: self-rating, the Hall-Jones definition, the Registrar-General's classification, the readership-based definition, and the multi-dimensional definition.

In the *self-rating* approach, the respondent is asked to select the social group to which he belongs from a list given to him; or he is asked to outline his own conception of the social structure and to place himself in it. Society is here conceived as being made up of psycho-social groups, which are essentially subjective in character and are dependent upon 'status consciousness'. The groups that emerge from the process may or may not conform to those which have been devised by survey researchers.

It is easier to outline the difficulties inherent in this approach than to know how to obtain its advantages without its disadvantages. The basic problem is that results are rarely comparable between respondents. That is to say, there is an in-built inconsistency in the method, since even if large numbers of respondents should state that they belong to, say, the 'middle class', there is no way of knowing if this term has the same (or similar) meaning for all respondents. The predictive value is, therefore, zero.

But the method does have advantages. For instance, suppose that various respondents' opinions of their status differ markedly from the status ascribed to them by some other ('objective') method. It is true, up to a point anyway, that people tend to act in a manner similar to that of the group with which they identify. Thus we may find that self-rating helps to explain the actions of groups that are either (I) deviant or (II) inadequately covered by an 'objective' rating, since their positions are changing so that they do not properly belong to any of the given, static, groups. It might be useful, for example, in identifying groups with strong aspirations to move upwards and whose 'life styles' are identified with those of the group to which they aspire.

The *Hall-Jones definition*[27] is worth considering in detail since reference has frequently been made to it in other studies. Three sets of different people were asked to make assessments of social groups, as follows:

I. 138 occupations were selected, and 5 people were asked to place them into 9 classes. (The 5 rankers were themselves of much the same class: they were a statistician, an officer of health, a psychologist, a woman graduate, and a member of the survey staff);

II. 30 selected occupations were ranked into 5 classes by 74 respondents, half of whom were attending WEA classes, and half of whom were people of similar occupation and education but not attending classes;

III. A wider enquiry was undertaken with the aid of some – mostly, white-collar – trade unions. This produced 1,056 returns in which the 30 selected occupations were again ranked in 5 classes.

In all cases the rankers were asked to assess the rank of the given occupations not according to their own assessment of their prestige but according to what they thought was the *general* social status rank order. At the same time Hall and Jones made up what they called a 'standard' classification of occupations, and compared the rankers' answers with this classification. No definition of groups into which the rankers were to place the occupations was given. No attempt was made to define 'social class'. The resultant groups were, therefore, a rank order, with the rankers given no indication of what was to be the nature of the groups. Having divided the occupations into the given number of groups, they were then asked to rank the occupations within each group.

Hall and Jones drew six main conclusions from their investigations. Firstly, the division between groups tended to be more an arbitrary division at convenient points in what should perhaps be regarded as a spectrum of status. Second, there was some overlap in judgements, there being greater agreement in judging the extreme groups rather than the central group. Third, by and large, there was a close consensus of opinion among the rankers. Fourth, the close correlation between the 'standard' classification and those given by the rankers suggests that there could well be a similar correlation between other 'standard' classifications and those given by rankers (for example, the Registrar-General's). Fifth, the differences in the rankings of the various occupations were greatest for those occupations in the centre of the spectrum, among all three sets of rankers. And, finally, the occupations of the rankers themselves appeared to affect their assessments of the occupations they were required to rank. In particular, manual and unskilled workers appeared to disagree widely in their rankings with those given by other rankers.

The main criticism of the Hall-Jones study is that it is, itself, unbalanced. Only 4 per cent of the rankers consisted of semi-skilled and unskilled workers, while 17 per cent consisted of professional and (high) administrative people. Two problems would seem to arise from this. In the first place, the ranking of occupations not included in the study is based upon the opinions of the few people who collaborated in ranking them, on the basis of (implicit and explicit) generalizations from the particular occupations covered in the study. In so far as the original study was biased by an over-emphasis upon professional people, so the ranking of the wider occupational groups will also be biased. And secondly, the under-representation of manual workers amongst the rankers in the original study appears to have caused some profound differences of opinion between manual and professional workers to have been unobserved. In order to explore this second problem further, Willmott and Young[76] carried out a small study among 82 manual workers in the East End of London.

Willmott and Young pointed out that although Hall and Jones did have some manual workers as rankers, these consisted of a disproportionate number who attended Workers Educational Association classes (who may be atypical). Their own study produced some significant contrasting conclusions. It confirmed that disagreements over the ranking of occupations was greatest among manual and unskilled workers. Indeed, they separated out 22 rankers whose rankings were so vastly different from Hall-Jones that they needed to be studied independently. But, even with the exclusion of these 'deviants' the gradings by the remaining rankers still did not agree closely with the Hall-Jones study.

The whole group of rankers in the Willmott-Young study tended to rate manual and skilled work higher than non-manual work. But, whereas the larger group tended to stress personal ability, education and money when making their rankings, the 'deviants' had equally reasoned (but different) bases for their judgements; they stressed the usefulness and importance of the particular job, and value to society. Willmott and Young stressed an association between occupational grading and political attitudes in their analysis of these differences; in other words the labour movement's particular and distinct ideology with regard to work. The labour movement has developed from and is still significantly based upon the craft-oriented trade unions, with their strong sense of the dignity and value of manual labour, and particularly a pride in (directly) productive work. The manual workers in the Willmott-Young study showed this clearly: they considered that their vital role in production gave them a high position in the rankings. Professional people, divorced from the labour movement and, often, with a tradition of opposition to it, tend to downgrade manual work and to consider that non-manual work is intrinsically more prestigious. Thus, there appears to be a tendency for people to be unable to differentiate between other groups of people that are a considerable social distance away from themselves.

This disagreement over status is very important. It should not be overlooked for the sake of creating a tidy rank order; nor because the social investigator (himself, a professional) may not have personal experience of the pride and consciousness of the manual worker. It means that 'social status' is much more complicated to determine than might, at first, appear to be the case.

The third basic method of grouping, the *Registrar-General's*, uses four combined bases for assessing social groups: occupation, industry, employment status and economic position. The object of the classification by *occupation* is to provide groups that have at least one common characteristic, that of work done. Where this creates groups which are too comprehensive, they are subdivided on the basis of other factors, such as skill, environmental conditions, and so on. Unit groups are obtained which are classified into 'orders' having certain broad occupational features in common. The classification by *industry* is simply based upon the definition by the Department of Employment and Productivity of standard industrial classes.[40] The classification by *employment status* is in two main divisions, each with subdivisions; self-employed (with or without employees, in large or small establishments) and employees (managers in large or small establishments, foremen and supervisors, manual and non-manual, apprentices and trainees, family workers, and employees not elsewhere classified). Finally, the classification by *economic position* is twofold: economically active and inactive persons. From the various groups produced in these four categories, the Registrar-

General[52] has constructed two sets of social groups; (I) social classes; and (II) socio-economic groups.

The five categories of *social class* are broad, and each is, as far as possible, homogeneous in relation to the basic criterion of 'general standing within the community' of the occupations concerned. This is naturally correlated with (and the application of the criterion conditioned by) other factors such as education and economic environment; but there is no direct relationship to the average levels of remuneration of particular occupations.

Each of the socio-economic groups was intended to embrace people whose social, cultural and recreational standards and behaviour are similar. The basis of the groups is, in practice, much more restricted than this, since there is a lack of data concerning these broader issues. In consequence, the classification is dependent upon groupings by employment status and occupation. There are 17 socio-economic groups, as shown in Table 7.

Table 7. *Registrar-General's Social Groupings*

A. Social Class
   I. Professional, etc. occupations
   II. Intermediate occupations
   III. Skilled occupations
   IV. Partly skilled occupations
   V. Unskilled occupations

B. Socio-Economic Groups
   I. Employers and managers in central and local government, industry commerce, etc. – large establishments
   II. Employers and managers in industry, commerce, etc. – small establishments
   III. Professional workers – self-employed
   IV. Professional workers – employees
   V. Intermediate non-manual workers
   VI. Junior non-manual workers
   VII. Personal service workers
   VIII. Foremen and supervisors – manual
   IX. Skilled manual workers
   X. Semi-skilled manual workers
   XI. Unskilled manual workers
   XII. Own account workers (other than professional)
   XIII. Farmers – employers and managers
   XIV. Farmers – own account
   XV. Agricultural workers
   XVI. Members of Armed Forces
   XVII. Occupation inadequately described

In practice, only two questions are necessary to establish the socio-economic group to which a person belongs. The first is concerned to identify the actual job which a person does; the second to discover whether or not he has any supervisory functions. Often, it is useful to begin with a third question concerning the firm or company for which he works and its type. This acts as a supplementary source of data to help identify the person's job in the event that his description of the latter is ambiguous or vague. An example of this series of questions, from the Birmingham Recreation Study,[65] is shown in Table 8.

With this information it is a relatively simple, though often tedious, task to identify the respondent's social class or socio-economic group by reference to the Registrar-General's classification.

The Registrar-General's groupings have the major advantage that they are based upon a comprehensive and detailed categorization of almost every possible occupation in Britain. They also have the advantage that they are widely accepted, and are, therefore, very useful for comparative purposes. They are based very strongly upon occupation, the data for which are usually easy to obtain. A dis-

Table 8. *Questions for the Identification of Social Class or Socio-Economic Group*

1(*a*)   What firm, company or organization do you work for?
PLEASE WRITE IN THE NAME

(*b*)   What type of firm is that? (For example, car manufacturer, tobacconist's shop, insurance office, etc.)
PLEASE WRITE IN TYPE OF FIRM

2   What job do you actually do there?
PLEASE DESCRIBE JOB IN YOUR OWN WORDS

3   Do you have anyone working under you? (For example, are you a manager, supervisor or foreman?)
YES
NO
IF YES
Approximately, how many people work under you?
PLEASE WRITE IN THE NUMBER

advantage is, however, that this reliance upon occupation may obscure other, more subtle, factors determining social status. Although there is clearly a close relationship between occupation and social status, we have no very clear indication of the strength of the association, or of the significance of other variables in this. Another difficulty is that the groupings are based upon the occupations of individuals, so that the social status of a particular household could fall within any one of several groups depending upon the occupation of the particular individual for whom data is obtained. Thus, for example, many working wives are employed in occupations which have a very different social status from that of their husbands.

The *readership-based* method of classifying groups has been used, though not extensively, by some market research organizations. It is based not on demographic factors, but on the reading habits of the respondents: the number and type of papers and magazines read. On the face of it, some interesting comparisons might be made between the social composition of, say, an upper-income suburban area and a working-class twilight area. But, although the method has interest, it also has grave defects. The main one is that readership is likely to be a dependent rather than an independent variable, related to such characteristics as education. Thus one might discover that *Guardian* readers have a strong preference for fishing and boating, but this would not tell one much more. Its value as a tool of analysis and prediction is not great. Also, any list of periodicals used would be sure to produce large, and not necessarily homogeneous, groups which read none of the periodicals listed. There would also be problems with changes in readership over time, and with the range of publications available at any point in time.

The *multi-dimensional definition* has been used, mainly by Warner[68] in the United States, but not as yet in Britain. Two or four status characteristics are selected, and each is given a weighting according to its (subjectively) presumed importance in the overall definition. Differences have arisen in the choice of characteristics and in the relative weightings given to each. Warner selected occupation, source of income, type of house, and dwelling area. Other rankers have selected education, possession of various consumer durables, and so on.

The main defect of the multi-dimensional definition is the degree of subjective appraisal that is required of the investigator: it is based upon a number of assumptions about status characteristics and their relative significance, for which there is no direct objective evidence. For instance, Warner's method begins by assuming a 7-class occupational scale. Although this is based, with modifications, on the

92

United States census, it is somewhat arbitrary. Anyone wishing to use this method would have to do a large amount of prior work on setting up an occupational scale. It is then necessary to make a weighting of the factors chosen. This is subjective and again raises problems. For instance, it is an assumption (and a questionable one), that occupation is twice as important as source of income in placing people on a scale; and any similar weighting assumption would be almost equally unreliable.

Having obtained a numerical scale for his respondents, Warner then divides them into groups on the basis of a procedure which he calls 'evaluated participation'. This involves asking families to place other families into categories on the basis of their known activities and status within the community. He then compares and studies the relationship between evaluated participation and a (derived) 'index of status characteristics', and from this lays out social demarcations. The method is highly subjective, depending as it does upon a community being small enough and closely integrated, so that people are able to rate each other. There also seems to be no particular rationale in going to the trouble of deriving a numerical index, only to divide it up on the basis of the subjective, and non-comparable, opinions of people about each other.

It is certainly true that Warner and Hollingshead, both using a form of the multi-dimensional method, achieved a large measure of agreement about social classes in Elmstown, USA, but this is not necessarily a validation of the method. This is because many classifications currently in use achieve a high measure of agreement, simply because, as we know already, there are many high and significant correlations between occupation, income, education, housing, and so on, and one is likely to come across this no matter what method one uses.

Another point is that, as Warner himself admits, his method has only slightly greater predictive power than if he used occupational groups alone; the extra work involved in the method may not be worth while. But the main defect is that the index is only evolved by comparison with existing and acceptable measures of social status; otherwise it is sheer guesswork.

One of the main conclusions that arises from a comparison of methods of classifying social groups is that there is a very strong resemblance between systems. As pointed out, there are so many correlations between factors that any reasonably well thought out system will produce much the same picture as any other one. This being so, it is essential to have a very clear idea of the kind of things one wishes to find out in a study in order to select the methods of finding them out. Practical considerations are that the groupings used should be based upon data that are reasonably easy to collect; that the less the need for subjective assessment and judgement on the part of the fieldworker the better; and that, where possible, the groupings should be readily comparable with those used for other studies.

All of this suggests that the most satisfactory approach would be to select a single characteristic as a basis for the first grouping and then to relate this to a number of other (presumed) relevant factors, one by one. For this, occupation has been chosen as the basic characteristic grouped by occupational units and then into socio-economic groups (or social classes) as done by the Registrar-General (Table 7). The reason for selecting occupation is that although it is impossible to state that such-and-such is *the* independent variable, occupation at least has strong claims to being a relatively independent one. Also, occupation based on the Registrar-General's tables has the virtue of producing results comparable with the Census; and of being based on a very comprehensive study of occupations. It is also a datum that is easy to collect, and one which requires no assessment by the fieldworker. In most cases, the socio-economic groups will be preferable to the social classes, since they give finer distinctions between groups.

Chapter 6

SAMPLES AND SAMPLING METHODS

Sampling is defined as the selection of part of an aggregate of material to represent the whole aggregate. All rigorous sampling requires the subdivision of the material to be sampled into sampling units. These may be natural units (for example, individuals) or aggregates of these (families and households) or artificial units (areas on maps). These units must be capable of clear and unambiguous definition. Clear definition demands the construction of a frame; that is, if households are to be the sampling unit, there must be available a list of households, and any that are selected must be capable of unambiguous location.

It is important in sampling to draw a clear distinction between a *population* and a *population of sampling units*. Any aggregate of values is termed a population. Thus, the population of sampling units consists of the whole aggregate of sampling units into which the material for survey is divided. If the sampling units are aggregates of the natural units of the material (for example, streets of houses), these natural units (that is, the houses) will form a further population which must be distinguished from the population of sampling units.

The major problem in sampling is, of course, to avoid bias. The latter can arise in two main ways; (I) through errors in the selection of the sample; and (II) through chance differences between the members of a population that are included within a sample and those that are not included. Bias arising from errors in selection forms a constant component or error which does not decrease as the size of the sample is increased. True sampling error – that is, error arising from chance differences – does decrease with increasing size of sample. Bias arising from errors in the selection of a sample is caused in several ways:

95

    I. by the deliberate selection of 'representative' samples;

    II. by unconscious cheating on the part of the investigator who allows his desire to obtain a particular (predetermined) result to affect his selection of the sample;

    III. by the substitution of an alternative convenient unit in the sample when difficulties are encountered in obtaining data from the originally selected unit;

    IV. by a failure to obtain data from the whole of the selected sample – that is, through nonresponse.

Much of this 'selection bias' can be avoided by a strict adherence to random sampling procedures – which, preferably, leave a minimum responsibility for choice of sample to the field staff.* Conversely, true sampling error – or random sampling error – can be reduced by various sampling procedures; and is, therefore, an important element in determining the choice between different sampling methods. It is doubtful however, that random sampling error can ever be totally eliminated. One final cause of bias which merits mention is that arising through faulty analysis of sample survey data – for example, by incorrect weighting of findings.

Sampling Methods

There is a wide range of different methods of sampling, none of which can be described as the 'correct' or 'best' method. Which is used will depend upon the particular requirements of each survey and upon the type and quantity of the population of sampling units. A brief outline is given below of some of the more important and useful methods. More detailed outlines of these are to be found in the two standard texts on the subject, by Yates,[77] from which most of what follows is drawn. Additionally, Kemsley[33] gives an account of the operations of the Government Social Survey, and the description of the sample design introduced in 1967 is instructive.

Random Sample. After the material to be sampled has been divided into sampling units, the required number of units are selected at random from the population of sampling units. Various tables of random sampling numbers can be used, but the procedure ensures that each sampling unit has an equal choice of being selected. There are various refinements of the workings of the method, designed to ease the actual process of choice but these do not affect its basis.

* It is worth noting, however, that it may not be necessary to obtain a complete absence of bias, if this can be shown to be (I) small overall and constant in time, and (II) trivial relative to true sampling error.

Stratification with Uniform Sampling Fraction. The population of sampling units is subdivided into groups, or strata, before the selection of the sample. These strata may contain the same numbers of units in each, or differing numbers. With the uniform sampling fraction method, the same proportion of units in each stratum is included in the final sample, these units being selected by random sampling procedures from the population of units in each stratum. This procedure has two purposes – to increase the accuracy of overall estimates, and to ensure that subdivisions of the sample population which are of interest are adequately represented in the final sample. Stratification affects random sampling error, since only variations within each stratum will give rise to sampling error. If a random sample is large enough, it will produce results which are virtually the same as those produced by a stratified sample; in particular, it will adequately cover subdivisions in the sample population.

Multiple Stratification. A sample population may be stratified for two or more characteristics. The population of sampling units is subdivided into groups or strata, which are again divided into subgroups or substrata; for example, towns can be divided into regions and then into population-size categories. The procedure is then the same as for ordinary stratification, each substratum being treated as the equivalent of a stratum. The purpose, is, of course, to further increase the accuracy of overall estimates and to ensure that more exact subdivisions of the sample population are adequately represented in the final sample.

Stratification with Variable Sampling Fraction. With some material, significant gains in accuracy can be obtained if differential sampling fractions are used for the various strata. This is particularly so where the stratification of the material has been carried out on the basis of 'sizes' of one kind or another. In such cases, the various quantitative characteristics of the units in the sample population often have within-strata deviations which are roughly proportional to the mean sizes of units in the different size categories. Then, sampling fractions should be about proportional to mean sizes.

Systematic Sampling from Lists. When a list is available of all units of the population to be sampled, the sample can be compiled by selecting every *n*th unit on the list, the value of *n* being determined by the size of the sample that is required. It is customary to select the first unit by the random choice of a number falling between 1 and *n*. Because of the lack of definition of strata, it is impossible to

G

make a very accurate estimate of the sampling error; but, provided there are no periodic (regularly recurring) features in the list, bias will not be significant. Systematic sampling from lists may often be more convenient to use than random sampling procedures. A commonly used list is the Electoral Register.

Multi-Stage Sampling. The population which is to be sampled is assumed to be made up of a number of first-stage sampling units, each of which is made up of a number of second-stage sampling units, and so on. The first-stage sample is obtained by a chosen (random) sampling method in the normal way; then, a second-stage sample is selected from each of the first-stage units by a sampling method which may be the same as, or different from, the method used to obtain the first-stage sample. The method introduces flexibility, enabling 'natural' divisions and subdivisions in the total population to be used as units at different stages of the sample determination. It permits, for example, the areal concentration of field work for surveys in which the total population to be surveyed is spread over large areas. In general, however, multi-stage sampling is less accurate than a sample containing the same number of final-stage units which has been selected by a suitable single-stage method.

Sampling with Probabilities Proportional to the Size of Unit. This method is particularly useful for surveys relating to areas of land of one kind or another. If areas of, say, fields are marked on maps and a point is randomly located on these maps, then the probability of this point falling within the boundaries of different fields is proportional to the areas of the fields. Thus areas can be selected with probabilities proportional to size simply by taking random points on maps.* The method may be applied to other kinds of material by forming successive subtotals of sizes of units, and, then, selecting numbers at random from the total of all units; but, in such situations, it would probably give greater accuracy if the material were stratified by size, and the sample then selected by means of the variable sampling fraction method.

Sampling from within Strata with Probabilities Proportional to Size of Unit. This method is mainly of use when units have been stratified according to some characteristic other than area, and when the number of units to be selected from each stratum is very small. It is particularly useful for second-stage sampling in a situation in

---

* It should be noted that an individual area may be included twice or more in the sample by this method. If this happens, the area must be counted each time so as not to distort probabilities.

which there are large first-stage units of variable size. The first-stage units are selected from within strata with probabilities directly proportional to size, while the second-stage units are selected with probabilities inversely proportional to the size of the first-stage units. In this way, the overall sampling fraction is kept constant, thereby simplifying computations.

Multi-Phase Sampling. It is sometimes convenient to collect certain data from all units of a sample, and other data from some of these units only, the latter being a subsample of units in the original sample. These samples may be chosen by normal (random) sampling procedures. Multi-phase sampling differs structually from multi-stage sampling. In the former, some sampling units are used throughout the survey, while, in the latter, a hierarchy of units is used.

Balanced Samples. If the average value of some quantitative characteristic of the units to be sampled is known for the whole population – for example, size – it is possible, provided that the sizes of the individual sampling units are known, to select a sample so that the average size of the selected units is equal to the average size of all units of the population. This is only satisfactory, however, if it is otherwise equivalent to a random sample. The method may be employed in conjunction with stratification – either for the whole population or for each stratum separately. Balancing for known quantitative characteristics is an alternative to stratification by size-groups for these characteristics. But it is only effective if the differences in the quantities under investigation are approximately proportional to the differences in the known characteristics; whereas stratification by size-groups takes account of any kinds of relationship, including non-proportional ones.

Systematic Sampling from Areas. This is a method of sampling material which is continuously distributed in space or time, by taking sampling units which are distributed at equal intervals over the material. Its chief application is in sampling land areas, usually by grid locations on maps. The method differs from *Systematic Sampling from Lists* in that it is a spatial methodology. Most lists do not correspond to any physical distribution, and, therefore, units from these are much closer to those obtained from random-sampling than are units obtained by systematic sampling from areas. But, provided that there are no periodic (regularly recurring) features in the material which is being sampled, it is a satisfactory method of area sampling.

99

Line Sampling. This is an alternative to point sampling for determining the proportions of a given area which are of different types. Sets of parallel lines, or strips, are taken as the sampling units, and all points along these lines are investigated.

Principle of the Moving Observer. This is used to assess the numbers of individuals moving about within a given area – for example, crowds in streets. Fixed-point counting does not take account of the velocity of movement up and down a street; nor, unless points of access are all manned, does it take account of movement in and out of the area. The moving observer traverses an area in one direction counting the numbers of persons he passes – in either direction – and deducting those who overtake him. He then repeats the process traversing the area in the opposite direction. The average of the two counts gives an estimate of the average number of people in the area during the time of the counts. It is valid provided only that all individuals can be readily counted and that the passage of the observer does not itself influence the movements of the individuals in the area.

Interpenetrating Samples. It is sometimes useful to take two or more samples of a given population, using the same sampling procedure for obtaining each. The two samples may be used for successive stages of a survey; or for obtaining separate and independent estimates of the characteristics of the population; or for comparing the results obtained by different investigators.

Sampling on Successive Occasions. If a population is subject to change it may be necessary to undertake successive surveys in one of the following ways:

    I. a sample survey may be repeated at intervals, a new sample being selected on each occasion without reference to previous samples;

    II. a sample survey may be repeated using the same sample as was originally used;

    III. a complete census or survey may be repeated at intervals in the original form;

    IV. part of the sample may be replaced on each occasion, the remainder being retained;

    V. a resurvey may be made of a subsample of the original sample.

Independent samples (I, III, and IV) are formally equivalent to interpenetrating samples. The fixed sample (II) is formally equivalent

100

to the observation of different characteristics of the same sample. The subsample procedure (v) is formally equivalent to a two-phase sample. The relative advantages of these various procedures depends upon the relationship between the variability of the sampling units and the variability of changes in these units.

Sampling Frames

The whole structure of a survey may be determined, to a considerable extent, by the nature of the frame from which the sample is drawn. Thus, for example, an intention to undertake a longitudinal study over time may be frustrated because of the lack of a suitable frame which includes persons moving house during the period of the study. Again, the structure of a survey (and the validity of the findings) may be affected by the degree of universality of the frame; that is, the frame may be incomplete. In fact, we can identify five main defects which are often found in sampling frames; frames may be (I) inaccurate, (II) incomplete, (III) subject to duplication, (IV) inadequate, (V) out of date. A frame is *inaccurate* if there is wrong information about the units comprising the frame – including the listing of units which do not exist. It is *incomplete* if some units are omitted entirely or if some information about some units is missing. It is *subject to duplication* if some units are included more than once. It is *inadequate* when it does not cover all the categories of data that are required for the survey to be mounted. And, finally, it is *out of date* if its composition is likely to have altered significantly between the time of its compilation and the time of its use for the survey.

These defects have different consequences for the validity of the survey. Inaccuracy will usually be discovered and corrected during the course of the survey, and therefore is unlikely to invalidate the findings. Incompleteness will rarely be discovered, and this usually means that units with some special characteristics will be seriously under-represented in the findings. Duplication has a similar effect giving some units a chance of double selection and representation; under certain circumstances, however, it may be detected. Inadequacy will often be known before the survey is begun – from the specifications of the frame itself. Generally this can be dealt with only by the construction of a subsidiary frame which covers the categories of required data that are missing from the primary frame. Most frames have these defects to some degree or other, so that the investigator will need to examine the frame carefully before undertaking the study, with a view to minimizing and 'balancing' these defects.

101

There are six main kinds of frames which are suitable for sampling human populations. They are, as given by Yates[77]:

    I. Lists of individuals in the population (or parts of it) which are compiled for administrative purposes;
    II. Aggregates of census returns from a complete census;
    III. Lists of households or dwellings in given areas;
    IV. Town plans;
    V. Maps of rural areas;
    VI. Lists of towns, villages and administrative areas, which often include supplementary information.

Lists of individuals as frames are only likely to be complete, accurate and up to date if the administrative machinery is very efficient and if there is some definite administrative need for the lists to be kept under constant revision. Lists of individuals are not suitable as a frame for sampling households unless the individuals are grouped by households. If addresses are selected from a list which is not grouped in this way, the probability of selection will be proportional to the size of the household.

Frames from complete population censuses have one major defect. A census, by its nature, is taken only at infrequent intervals. It is, therefore, out of date for a large part of its life. This problem can be overcome if, at the time of the census, a master sample is constructed, from which further samples can be drawn as required. The sampling unit in the construction of the master sample should be the dwelling, not the individual or household occupying it, as this will have greater permanence.

Lists of households or dwellings are frequently available from such sources as rating officers. Frames based upon lists of dwellings are more permanent than those based upon lists of households and individuals; since, if complete at the time of their construction, they will only become incomplete as a result of new building or changes in the uses of existing buildings. Lists of households can be used to construct a frame based upon dwellings by taking as the sampling units the dwellings occupied by the households at the time the list was constructed. A procedure known as the 'half-open interval' may be used to include dwellings which were non-existent or unoccupied at the time the list was constructed. When drawing the sample, the dwelling unit appearing next in the list to the one which is selected for sampling is noted. The interviewer is then instructed to see if there is any other dwelling on the ground between the dwelling unit selected for sample and the one which is next to it on the original list. If so, the intervening dwelling is to be included in the sample. Obviously this technique is only possible if the original list corresponds to some geographical pattern on the ground. (It may, of course, be possible to rearrange the list to give such a pattern.)

Frames from town plans are often useful and relatively simple to operate. If a town plan gives an accurate representation of streets, than the town can be divided into 'blocks' bounded by streets. A sample of dwellings can then be drawn from these blocks. Indeed, the blocks can usually be taken as a first-stage sample and dwellings as a second-stage thereby reducing the amount of work that needs to be done, since lists of dwellings need to be drawn up only for those blocks selected in the first-stage sampling and not for all blocks. If, however, the numbers of dwellings in each block vary significantly, it may be necessary to stratify the blocks.

The remaining two categories are frames from maps of rural areas and from lists of villages and administrative areas. If there are available maps which show all, or virtually all, the buildings in the given areas, then sampling can be undertaken by constructing rectangular areas. It may be better, however, to divide the maps into areas containing approximately equal numbers of dwellings, using natural boundaries as far as possible, and then sample from these. Villages of varying sizes falling within these rural areas are best dealt with separately – by some form of stratification and two-stage sampling. Where the population of a rural area is concentrated into villages and accurate maps are not available, sampling – by stratification according to, say, population size – may be by lists of villages. Finally, aerial photographs have been found to be particularly useful for land surveys. They are not, strictly speaking, frames, since they provide the actual data of the survey. The handling of aerial photographs is a highly skilled operation, since they are subject to distortion resulting from variations in the height from which they are taken and from the tilt of the camera – although there are technical innovations in use which have reduced the scale of this problem.

Sample Size

Having decided upon the type of sampling method to use and the sampling frame that will be employed, there remains the question of sample size. The size of the sample required to achieve accuracy depends upon the variability of the material and upon the extent to which it is possible to eliminate the different components of this variability from the sampling error. The determination of the size of the sample is based upon the *standard error*. This is a measure of the average magnitude of the random sampling error that can be expected from any given random sample. The formula for the determination of the value of the standard error is quite simple. If p is the proportion of units of a given type in the whole population and $q(= 1 - p)$ is the proportion which is not of this given type, the standard error of the proportion of units of the given type that will appear in a random sample of n units is given by the formula $\sqrt{(pq/n)}$. The formula also holds if the proportions are replaced by percentages, i.e. $\sqrt{\{(p\%)(q\%)/n\}}$. If, for example, 20 per cent of the units of the population are of the given type, the standard error of the percentage units in a sample of 1000 is $\sqrt{(20 \times 80/1000)} = 1\cdot26$. This formula, rewritten can then be used to give the number (n) of units required in a sample when the required standard error is known, i.e. $n = pq/(\text{required standard error})^2$. Thus, if a sample is being taken of a population in which it is believed that about 20 per cent of the units are of a given type, and it is required to determine this percentage with a standard error of 1 per cent, the size of the required sample will be as follows: $n = 20 \times 80/1^2 = 1600.$* These formulae hold only for a random sample in which the sampling units are the units for which the proportion having a given attribute require to be estimated. In sampling a human population, for example, the sampling unit may be the household and not the individual. In this case, the standard errors of proportions of individuals determined from the sample will be larger than those given by putting n equal to the number of individuals in the above formulae.

The calculation of the size of sample required to obtain a given degree of accuracy is relatively simple in the case of a random sample. With more involved types of sampling, the calculation is more complicated. There are, however, some general rules which should be observed:

I. A calculation of the accuracy that would be obtained by means of a random sample is a useful preliminary guide to the size of sample that is likely to be necessary.

II. The use of stratification, a variable sampling fraction or supplementary information will, in general, increase the degree of accuracy obtained. This increase will generally be larger for quantitative characteristics of the population than for qualitative ones (that is, attributes).

III. Stratification will only increase the accuracy substantially,

* There is, also, the concept of the 'percentage standard error' which can be utilized as alternative to the 'standard error'.

however, if there are marked differences between the different strata of the sample.

iv. A variable sampling fraction can greatly increase the accuracy when the sampling units vary greatly in size, or more generally in variability from stratum to stratum.

v. Since there must be at least one unit per stratum, more detailed stratification is possible with larger samples.

These rules are indicative only. The final decision as to the type of sampling to be adopted necessarily depends upon the relative accuracy of the various methods and their relative costs.

Decisions on many of the points that have been outlined above will often be possible only after preliminary studies in the form of pilot surveys have been undertaken. The objects of a pilot survey are threefold: to provide information on the various components of variability to which the material is subject; to test questionnaires and to train field workers; and to determine the most appropriate and effective type and size of sampling unit.

A fully random sample is not necessarily the most appropriate for pilot surveys. For example, it may appear that a given survey is most likely to be appropriately covered by means of a stratified sample. If so, it will be necessary, by means of a pilot survey, to determine the components of variation within strata. This can only be done by a fully random sample (for the pilot survey) if the sample is sufficiently large relative to the number of strata for the majority of the strata to contain at least two units. This difficulty can be overcome by adopting some form of multi-stage sampling, so that the whole of the pilot survey can be concentrated in a few of the strata. But then, multi-stage sampling creates other problems for piloting. The component of variability that governs the sampling error for the whole population is the variability of the first-stage units. The pilot survey must give an adequate representation of these, selected at random, to make a reliable estimate of variability. Thus a more extensive pilot survey is required for multi-stage sampling than for a single-stage sample, if the sampling error is to be estimated reliably.

Quota Sampling

Finally, consideration needs to be given to one of the most controversial issues in survey sampling at the present time; quota sampling. In quota sampling, interviewers are given definite quotas of people in different social classes, or different age-groups, and so on, and are instructed to obtain the required numbers of interviews in each quota. Additional instructions, designed to prevent a significantly unrepresentative selection within the allotted quotas, may also be given, relating to such items as method of contact.

In general, academic statisticians criticize quota sampling for its theoretical weaknesses, while market research investigators defend it for its cheapness and ease of practical application. There are six main arguments against it. It has been argued that it does not meet the requirements of random sampling and, hence, standard errors cannot be calculated for quota sample results. Again, quota samples invariably utilize socio-economic variables for control, but socio-economic classifications are based upon a fragile statistical foundation, and, as controls, they are inevitably defined in vague terms. Moreover, bias may be introduced within quotas because interviewers may not obtain a representative sample of respondents in these quota groups. Another argument is that bias may arise through the peculiarity of the interview situation – for example, in streets, offices and factories. Perhaps the most telling argument, however, is that not only has the interviewer too much freedom in choosing respondents, but the method also allows too little control of the field-workers by the investigator. It is, for example, more difficult to check on the honesty of field-workers in a quota sample than in a random sample. Finally, quota samples disguise the influence of refusals, which may thereby cause bias.

Against this, it is argued that random samples are not necessarily perfect owing to refusals and non-contacts, and so the calculation of the standard error is for a non-perfect final sample anyway. Moreover, the sampling error is of small importance compared with the considerable and intractable non-sampling errors arising in the collection of the data. This negative argument is backed up by more positive ones; for example, that quota sampling is very cheap, arising, in part, from the lack of call-backs. Again quota sampling is easier to operate from the administrative point of view. There is no stage of drawing a sample, no problems arising from non-contacts, no refusals, and so on. Moreover, if the fieldwork has to be carried out in a very short period of time, quota sampling is usually the only feasible method. Also, since there are only a few, broad social groups usually employed, disagreement about the correct classes of respondents are likely to be rare. Again, quota sampling does not require the existence of frames. Finally, even when quota sampling is recognized to lead to bias, this can often be identified as being related to certain characteristics only, and should not affect the validity of findings related to other characteristics.

Moser and Stuart[43] tested some of these conflicting claims through a number of studies in 1951 and 1952. The studies employed three main methods; the analysis of data from past surveys; the analysis of demographic data from two national quota and random surveys; and the results from four field experiments.

Comparisons based upon the analysis of data from past surveys were of very limited value, but were useful in one respect; a test of differences between geographical spreads of samples and population density (by chi-squared) suggested a very marked 'bunching' of quota samples.

The comparisons of demographic data from the two national quota and random sample surveys, both carried out in Britain in 1951, showed that, on most questions, there were few significant differences. Age and sex distributions were very similar – as were household compositions, by both size and distribution. On questions related to social and economic status, however, the two samples differed markedly. The percentage of 'refusals' and 'don't knows' was twice as high for the quota sample as for the random one. There were large differences, by occupation and income, among those who did supply the data. This bears out the contention that quota samples usually produce an 'industrial' bias in occupational data. The distribution in the random sample corresponded closely to the correct (that is, 1951 census) distribution.

The main results from the field experiments were that the refusal rates for the quota samples were more than twice the figure for the random samples; that there was a general tendency for quota samples to over-represent the lowest ages with control age-groups and to under-represent the highest ages; that quota samples under-represented workers in manufacturing industries, while over-representing those in distributive, building and transport industries; and that there was a pronounced tendency for the quota samples to under-represent persons who finished their education before the age of 15 years. (The random samples, however, corresponded closely to the Census figures.) In addition, the quota samples produced a more accurate distribution by marital status than did the random samples. There were also sizeable differences in the household size distribution between the random and quota samples. In particular, there was a higher proportion of respondents from 1-person households in the quota samples. Finally, variability in the quota samples was well in excess of the random samples. Generally, though, the authors concluded that there were few *major* differences between the results of quota samples and those of random samples. Age, sex, and social class controls, however crude and subjective, do much to bring different quota results into line.

These conclusions do not, however, establish the theoretical

107

soundness of quota sampling. The method is not suitable for surveys in which it is important that results are derived (and known to be derived) from theoretically safe sampling methods. Against this, it should be noted that nonresponse constitutes a practical weakness in theoretically sound random sampling methods. But then, again, nonresponse may be just as significant, but hidden, in quota samples. Essentially, it can be claimed only that quota sampling worked satisfactorily with skilful operators. (Even then, the experiments were not totally satisfactory; they were rather too complex for their size, giving many tentative conclusions rather than a few definite ones.)

This brief discussion of sampling methods provides an outline of some of the general rules that need to be followed in conducting sample surveys. It does not purport, however, to give a detailed analysis of different kinds of sampling methods and frames. It is written for the non-specialist research investigator with little or no statistical expertise but with a need to know the broad procedures and problems associated with sampling. More detailed consideration of most of the issues discussed here can be found in Yates[77] and Kish,[35] the two major texts on the subject. The former is British and probably more useful, therefore, for research workers in this country, especially in respect of sampling frames.

Chapter 7

DATA ANALYSIS

This chapter provides a brief description of some of the equipment available for data analysis, and makes some general proposals about the organization of the work of analysis. It consists, largely, of notes on the *operational* details of data analysis, and not on the *interpretative* analysis by the application of various kinds of statistical tests. There is very little that has been written, in comprehensive form, about the operational aspects of data analysis and, in particular, the various machines by which analysis can be carried out. The outstanding work is by Yates,[77] upon which much of what follows is based. Other sources are Moser[42] and the Survey Research Center, University of Michigan.[60]

The steps to be taken in data analysis are broadly sevenfold:

I. Preliminary computations of the value of individual returns such as the calculation of ratios and the introduction of weighting factors (if punched cards are used, some of this can be done mechanically subsequent to the punching of the cards).
II. abstraction or coding of the results so that they are in a form suitable for analysis or for transfer to punched cards;
III. punching of the cards;
IV. counts and totals;
V. preparation of summary tables from the counts and totals, including adjustment for supplementary information and any weighting not yet carried out;
VI. calculation of sampling errors and investigations of efficiency;
VII. critical analysis.

At an *operational* level, there are five main ways of handling the data: analysis direct from the forms; transfer of the data to plain cards; the use of Cope-Chat cards; the use of punched cards (that is Hollerith cards); and the use of electronic computers. A few notes, on the use of electronic computers are given later in this chapter; but, at this stage, we can safely say that the analyses performed by punched-card machinery can be performed with greater speed by electronic computers. These computers can obviously also carry out many more sophisticated forms of analysis. For the present, however, remarks are restricted to the more common and simpler forms of analysis and to ordinary punched-card machinery.

The sorting of plain cards must be entirely by hand, and is, therefore, very time-consuming; Cope-Chat cards merely aid hand-sorting processes; while, with punched cards sorting is entirely mechanical. In some cases, data can be recorded directly on to the cards used in analysis, thus eliminating the coding process. But only small amounts of uncoded data can be put directly on to ordinary or Cope-Chat cards, and such cards tend to become damaged during the field work. With punched cards, a process known as 'mark sensing' is now used for some surveys (for example, in the USA, for gas and electric meter reading). For this, the interviewer (or respondent or meter reader), uses a card similar to a standard IBM card, and with the electrographic pencil blackens parts of the card, each part representing a particular piece of information. The card is later fed through a reproducer which senses the marks and translates them into punched holes on an ordinary IBM card. The machine also makes checks for inconsistencies in the data. This method, it has been suggested, can save up to two weeks in processing. It is, however, more suitable for surveys where fairly standardized information is required of large numbers of persons. Where more complicated information is sought, such as responses to non-standardized questions, mark sensing is of relatively limited value. It is also probably better used when the interviewer is to fill in the cards, as problems arise in explaining the method to respondents and getting them to use it without errors.

It is worth considering hand sorting or analysis direct from the forms only for small-scale surveys. If the data are to be transferred to cards, then coding will be necessary. If Cope-Chat cards are used, all the information to be recorded in punched form will have to be coded; and if punched cards are used, then *all* of the information will be coded. (The difference is that Cope-Chat cards have a space in the centre of the card in which information can be written.) The disadvantage of punching all the information is that detailed data relating to separate respondents often cannot be transferred to the cards and, therefore, examination of some items in depth becomes difficult. Where the size of the survey necessitates the use of punched cards, it is advisable, therefore, to keep the original forms so that detailed investigation of particular matters may later be carried out.

Cope-Chat Cards have holes along each edge. Each hole represents one group in the classification of responses to a given question. These cards are particularly useful for quick sorting, since a knitting needle or similar instrument can be passed through a particular hole and the other cards shaken out. Cards can thus be sorted into classes with more speed than by hand. Once the cards are sorted, however, they must still be hand counted. Isolated cards with given characteristics can be extracted much more readily with this method than with plain cards. There is also an advantage over punched cards in that no elaborate sorting mechanism is required. The volume of data that can be included on each card is, however, relatively limited, since the usual size is around 5 inches by 8 inches with four holes to the inch, giving 105 holes in all.

Various types of punched cards are available. It is best to get advice from the company concerned about the use of the chosen machine. The number of columns available varies, but the standard IBM card has 80 columns and 12 rows. Sometimes each column can be made to contain two items of information, by a system of multiple punching, thus doubling the capacity of the card. Punching is done by a hand-operated key punch; a hand-operated verifier, similar to the key punch, being used to check that the punching is correct.

In general, expert advice should always be sought in planning extensive punched-card work, *before* coding is carrried out, and preferably at the time when the questionnaires are being drawn up. This is because the work of field staff, coders and punch operators cannot be considered separately – each group of people must use the forms or schedules, and the convenience of each should be considered. Also, the whole process of coding and analysis must be planned as a coordinated operation. If expert opinion is consulted at an early stage the chances of collecting useless data, or data that cannot be analysed, will be reduced.

There are several mechanical aids to data analysis, with varying degrees of refinement. For present purposes, however, we can distinguish six types of equipment; the sorter and sorter-counter; the tabulator; the reproducing summary punch; the multiplying punch; the collator; and the computer. A brief description of each, taken from Yates[77], is given below, but it is not necessary for the research worker himself to have more than a general knowledge of each and its function.

The sorter scans one column of a card at a time. All cards are passed through the machine and separated into twelve boxes corresponding to the twelve possible positions of the holes punched in the column. Thus if a column refers to one question with twelve possible responses, the cards will be sorted in accordance with which of the 12 holes in the column has been punched. The sorter is generally used to sort cards before passing them through the tabulator. Some sorters have a counting device, which counts the number of holes occupying the various positions in a given column.

The tabulator is a much more elaborate machine than the sorter. Its primary function is to add the numbers punched in a given column from a set of cards. Counters accumulate the numbers as the cards are passed through and the printer prints the totals. More elaborate tabulators have various auxiliary devices, the most important of which are the 'rolling feature' and the 'distributor'. The rolling total tabulator adds or subtracts numbers from one counter to another. Rolling can also be used for simple multiplications. The distributor enables numbers read from a single column to be directed to different counters, and also enables numbers taken from one counter to be directed to other ones for rolling.

The advantage of the tabulator over the sorter-counter is, of course, that it prints its results. Also, the whole pack of cards is kept together. But the sorter-counter is generally better for counting when there is multiple punching in a column.

The reproducing summary punch has two main functions: firstly, it can reproduce the information given on one pack of cards on to the corresponding cards of another pack; and secondly, it can do gang-punching – that is, the punching of information read from a master card on to a whole group of cards. It can also be used with the tabulator to punch on to new cards the results obtained in tabulations. Thus, for example, the punch can be used to make a new pack of cards when the old ones are worn out. It will bring together on a single card items of information drawn from separate cards but which refer to the same respondent, so that the relationship between different items can be analysed. New information can be put on to cards that are already punched, and information from one pack may be put on to another pack already containing information. The machine will check that the two packs correspond, and will check that the transferred information is correctly punched.

The Multiplying punch will read two numbers from a card, calculate the product and punch the result (with suitable rounding off). It will also multiply a whole group of cards by a factor given on the master card. It will add and substract two or three numbers, or add one or two numbers to the product of other numbers. It is used primarily for the calculation of various products prior to summation and for the calculation of ratios. It is, however, rather slow in operation.

The collator will take two packs of cards previously sorted in numerical order on a number of columns and combine them into a single pack, so arranged that all cards from the second pack which carry a given code number will follow on cards from the first pack with the same number. It will select matching cards from the two packs. The collator can also be fitted with a card counting device which picks out every *n*th card. This is helpful in extensive systematic sampling from a pack of cards.

Electronic Computers will carry out all of the activities described for punched-card machinery. They work at a considerably higher speed, and will carry out automatically various checks and tests. Computers have a large memory store for holding numbers. For large-scale processing work where greater storage capacity is needed, this is done on magnetic tape. There are several alternative methods of storage which involve varying speeds of access. With an electronic computer there is no need to adopt an elaborate coding system for the input of information, since the data can be interpreted and coded by the computer as it is fed in. This greatly facilitates the punching stage, as the information contained in the basic schedules, provided it is reasonably concise can be punched in the form and the order in which it is recorded. Input is in the form of punched cards of standard type, or paper tape. Cards are read, row by row, sometimes simultaneously; as each row is read, the data recorded are processed according to the input orders.

The computer is capable of carrying out a wide range of computations – both simple and elaborate. At one end of the scale it will compute standard errors; at the other, it will carry out elaborate analyses in examining crude tables of class means, percentages, and so on, to throw light on underlying relationships and the effects of various factors. But critical analyses of the more complicated kind will usually only be carried out after the main tabulations have been examined, and it has been decided which analyses are both required and worth while.

Obviously, the high speed and accuracy, as well as the potential sophistication of the computer are its important attributes. But proper organization of the work to be done is even more necessary than with punched-card machinery, if computations are to be

H

effectively and economically carried out. Before a start can be made on programming, decisions must be taken on the required preliminary tabulations, editing, and similar details. Final decisions can again often only be reached at a later stage, after preliminary analysis of the data has been made, in conjunction with a careful study of pilot results. At all stages of planning, expert advice should be sought on methods and systems. More detailed information about all of these mechanical aids to analysis is to be found in Yates.[77]

Attention can now be directed away from mechanical methods of handling survey data to questions of coding, computation and accuracy. All non-numerical information collected in a survey must be coded for punched-card analysis, in quasi-numerical form, (for example, X, Y, 0, 1, ... 9) giving twelve classes to each column. Numerical information does not require coding, but often needs rounding-off to economize card space and reduce the counter capacity needed. Rounding-off, as a separate operation, can often be avoided by appropriate instructions to field-workers. Sometimes however, it is helpful to code numerical quantities, especially when a quantity is required primarily as a basis for nominal groupings and not for precise summation. If precise totals are needed, however, then large values must not be too closely grouped and there must be no open category – that is, no 'over x' group.

Usually three sets of people must use the questionnaire or schedule – the field workers, the coders, and the key-punch operators. The field workers will be dealing with each document for several hours, the coders for several minutes, and the key-punchers for several seconds. Delays in handling the document, caused by bad design, will affect each group's work accordingly. Although there are some general requirements which are the same for all groups, the type of survey may greatly affect the difficulty of the work of one group rather than the others, and a choice has to be made by the survey director. Broadly, all three groups require that the document be clear; that questions (and their answers) be easily distinguished at a glance; that the order in which questions are to be asked or punched must be obvious; that the questions that are to be asked only of some respondents must be clearly marked; and that the places where interviewers are to record answers, where coders are to code and from which codes are to be punched must stand out and be separated from the rest of the schedule.

The complexity of the coding operation will greatly vary. If it is thought that two-way questions, or multiple-answer questions on a checklist are adequate for the purpose of the survey, such responses are easily coded – and can, indeed, often be precoded – thus considerably reducing the work of the coders. But these types of questions are not necessarily suited to all surveys, particularly where a deeper probing into respondents' opinions and feelings is required. In this case, free-answer questions are frequently chosen, as these do not restrict the respondent in any way, or suggest answers to him, but permit him to express freely and fully his own views, at any length and taking any viewpoint he wants. Often, no attempt is made to code responses to free-answer questions: they are asked only of a sample of respondents, or only a few questions are posed in this way, so that the processing and analysing of these responses by hand will not be too great a task. But the Survey Research Center, at the University of Michigan,[59] has developed a technique of coding for large-scale surveys where nearly all of the questions are free answer. Questions are designed to elicit a free and open response, and are generally followed by a neutral probe (for example, 'why do you feel this way?'). They have developed a complex coding system which is then applied to the results, and which attempts, by the use of skilled coders following detailed instructions, to reduce the mass of material to analysable form. The rationale behind this is good: it attempts to sort the material after it has been freely expressed and collected, rather than attempting to sort people into groups before they answer questions. The method by no means obviates the need for pretesting; but rather, the reverse. A code system has to be developed for each survey, and pretesting is needed to develop the system so that there are sufficient codes for most of the frequently-recurring types of responses. The method involves a very much more complex coding operation than the ordinary 'two-way' or 'multiple-answer' type of question.

Other general points about coding are that provision must be made for coding the absence of information as well as its presence. Standard codes must be included for 'Don't know', 'No answer' and 'Does not apply'. If 'Does not apply' is not coded the machine will have to return to the previous question, or the previous column, to sort out those with the open column that did not answer but should have done from those that did not answer because they should not have done so. Blank columns lead to confusion and should be avoided. Numbers below 100 should also begin with 0, not with a blank; that is, 092 not 92.

In general, compactness of coding helps both to save card space and to simplify sorting and grouping cards into classes. Two-column codes are usually more complicated than one-column ones for sorting and counting; separately-coded classes cannot be grouped for

115

tabulation unless they are gang-punched or the control omitted. It often pays, therefore, to code an item in two parts, main and subsidiary. For example, instead of numbering counties in Britain, it might be better to number the regions and then, consecutively, number the counties within the regions.

If there are many questions having only 2 or 3 answers, columns used can be reduced by combining answers to two or more questions in a single code, that is, two questions with answers 'yes', 'no', and 'don't know', can be coded in one column using nine combinations of answers. But this system can be troublesome, and is often less convenient for analysis. In large scale surveys it may be better to keep answers separate, even if this involves using another card. The same goes for multiple-punching, which, if used too much slows punching, is difficult to verify and complicates sorting and tabulation. It is better to use an additional card, and to recombine data as necessary using the reproducing punch.

The arrangement of information on punched cards depends on the analysis to be done, and on the type of survey. Sometimes, for reasons of rapport, questions which should be punched together are widely separated on the actual questionnaire. The problem is more complicated if the sampling units are not the units of analysis (for example, households may be the sampling units, but analysis may be of both individuals and households). But, generally, if the sampling units are the natural units of the population, information can simply be arranged in the order most convenient for punching. It is also best to leave a few blank columns, usually grouped at one end of the card for the punching of identification data.

After a coding system and the arrangement of punched information have been determined, the next step is to plan the computations. The appropriate methods of computation depend on the type of materials, the scale of the survey, the depth of analysis required and the available personnel and equipment. In certain circumstances, preliminary computations can significantly reduce the amount of work to be done on punched card machines: for instance, if only totals are required in analysis, it is better to obtain these before punching, and punch only the results on the cards. The question of economizing mechanical aspects of analysis is very technical and expert advice should be sought.

There is, often, a danger of over-mechanization in the analysis of survey material. The use of punched cards can lead to stereotyped and uncritical analysis. For investigational (as opposed to census-type or descriptive) surveys provision should always be made for further tabulations.

If there are defects in the sample or in the collection of data the different classes of the population are found to be represented in incorrect proportions in the final sample. Adjustments by weighting can reduce the scale of the problem but should not be expected to eliminate it. In general, however, it is better, if a sample is markedly defective, to give the unadjusted results, plus information which shows as far as possible the deviations from the expected distributions. Any adjustment made should be clearly stated and the magnitude indicated.

Nonresponse is a frequent cause of defects in the sample. The simplest method of adjustment is to treat nonrespondents as being similar to the rest of the sample; that is, to treat the sample as though it were merely a smaller one. This is unsatisfactory, however, when it is known that, or suspected that, nonrespondents are not similar to respondents. Alternatively then, if a follow-up is used and there is a reasonable response to the follow-up, the initial nonrespondents who subsequently respond are treated as a subsample of the initial nonrespondents and are weighted accordingly. The advantage here is that if there is a difference between respondents and non-respondents, clearly, follow-up respondents will be more like the initial non-respondents than the initial respondents.*

There is a difference between the types of deduction which can be made with certainty from survey data and those that are probabilistic or merely possible. Correlation does not prove causation. Even where a casual relationship is known to exist (such as between overcrowding and aspects of social malaise), deductions as to the magnitude of given factors can never be made from survey data (the households overcrowded may have had quite different social backgrounds and the individuals concerned may have quite dissimilar personality determinants.)[9] To determine, with certainty, the magnitude, *in a casual sense*, of any given factor, controlled experiments must be undertaken. Surveys are not substitutes for such experiments. But if survey data are to be effectively used, it is necessary to have means of eliminating the effects of extraneous factors, by weighting. Yates[77] outlines concisely a number of different ways of doing this.

Accuracy in analysis is extremely important. The attainment of a high standard of accuracy needs careful organization of the checking procedures and scrupulous attention to detail. Numerical work can be checked by such methods as repetition of questions, cross-checks, or using different methods of computation to achieve the same result. Even these processes can be subject to error; there is always a risk of

* See above, Chapter 3, page 38.

117

repeated errors, and so on. Cross-checks are helpful in eliminating errors in total counts which are based upon counts relating to each of the classifications. Even though it may not be possible to eliminate all errors (except where use is made of electronic computers, which have in-built checking mechanisms and where a very high standard may be expected), a checking system should be employed. If more than a few errors are found, all computations should be redone. Final results must be scrutinized for apparent inconsistencies and anomalies, and those found should be investigated. Sample checks on punching and coding should be made.

Different stages of analysis and different types of work demand different standards of accuracy. The main point about accuracy is that a procedure for achieving it at each stage must be devised, the need for which may only become apparent after the results of the first tabulations have been examined in detail. Flexibility should therefore be a prime requirement in planning the analysis. It may be possible to lay down the broad lines of future analysis at an early stage; but, especially in investigational surveys, the preliminary results nearly always lead to requirements that could not have been foreseen. The purpose of an investigational survey is, indeed, to do exactly this. In general, items of information should not be omitted from coding merely on the grounds that they are not required in the primary analysis. The broad aim should be to summarize all of the available information in coded form, so that, if new needs arise or further analysis is to be undertaken, the work can be done without recoding.

Close co-operation is needed between the researcher, the coding section and the machine section, if appropriate techniques are to be applied to the inquiry; also, between those producing the tables and those carrying out the statistical computations afterwards. The supervisory staff of the statistical work must ensure a high level of accuracy and must be able to reduce the most complex statistical calculation to a routine computational base. Some suggestions for the analysis and interpretation of survey data can be obtained from a study by the Government Social Survey.[24]

It is worth while to conclude this chapter with some brief notes about the compilation and presentation of reports giving the results of data analysis and surveys generally. Detailed guidance has been laid down in a manual prepared for the United Nations,[62] but some general rules can be cited here. Broadly, there are six matters to be covered: a general description of the survey, its purpose, coverage, timing, accuracy, and so on; then, the design of the survey in detail; followed by a description of the method or methods of sampling; then, an outline of the personnel and equipment used in the survey;

118

then, a discussion of costs, under such headings as pilot studies, fieldwork, processing and analysis; and finally a description and discussion of the findings of the survey, and, in particular, matters affecting the probable accuracy of these findings. A fuller outline of the recommendations of the United Nations report is to be found in both Yates[77] and Moser.[42]

Chapter 8

NON-SURVEY RESEARCH TECHNIQUES

There are other research techniques, besides the social survey, which can be employed as a means of establishing data about the social characteristics of given populations. Many of these are in early stages of development and their value is relatively limited. They can often be useful, however, as complementary or supplementary approaches to the social survey. They will be considered here under four main headings: physical evidence; mechanical and electronic devices; documentary sources; and observation.

Physical Evidence

Physical evidence refers to data which have not been collected specifically for research purposes, but which are available to be exploited by the research investigator. Webb *et al.*[70] distinguish two main kinds of physical evidence: *erosion*, where the degree of wear and tear on a given material serves as a measure; and *accretion*, where the evidence relates to a deposit of material. An example of an erosion measure is given by the experience of the Chicago Museum of Science and Industry, quoted by Webb. Officials noted, from maintenance records, that the vinyl tiles around an exhibit containing live hatching chicks needed replacing about every six weeks, while in the remainder of the museum they often did not need replacing over periods of several years. It was suggested that this provided a crude measure of the popularity of this exhibit as compared with others. Indeed, it could be hypothesized that the rate of tile replacement around the various exhibits would provide a crude index of the relative popularity of the different exhibits. This hypothesis was, in fact, tested – and partially confirmed – by direct observation. Thus, the measure of popularity was the amount of physical erosion. (The source was, however, maintenance records.)

Webb *et al.* have also provided an example of a physical accretion measure in their description of Sawyer's study of the sales of alcoholic drinks in Wellesley, Massachusetts.[54] Because of particular local difficulties, the usual methods of observation, survey and study of sales records, were not available to him. Instead, he examined garbage which had been collected from different parts of the town and counted the numbers of empty bottles. The resulting data, while incomplete and subject to unknown variables – for example, he had no knowledge of bottles which had been broken nor of those which had been retained in households for some purpose or other – nevertheless gave a partial measure of the incidence of alcohol consumption in the town.

The chief value of data derived from physical evidence is, in fact, for securing measures of incidence, frequency, attendance and the like. The major advantage is that physical evidence is entirely objective in character (although the investigators' interpretation of the evidence may introduce a subjective element into the study). The data are produced without the subjects knowing that they will be used for research purposes. This avoids the problems that arise from the bias of respondents in the more usual survey research methods. On the other hand, there is the major disadvantage that the quality of physical evidence may vary haphazardly over time and place. Some materials, for example, will endure for much longer periods of time than others – a problem which has been apparent in archaeological research for many years. Again, the coverage of physical evidence also tends to vary haphazardly. There is usually no possibility of systematic sampling of physical evidence; and there is, therefore, no clear indication of how representative the data are.

The major research value of physical evidence appears to be as complementary or supplementary information to data obtained by alternative methods of investigation; for example, by sample survey or observation. If physical evidence is used in conjunction with such methods, it will often be possible to control the problems arising from haphazard variations in the quality and coverage of data, while allowing some check on the degree of error arising from such factors as respondent bias. In planning, for example, it would be possible to utilize the degree of grass erosion on playing areas in parks, or the volume of refuse collection from open spaces, as measures of the volume of use made of such facilities. Together with records of matches played on sports areas, they could provide a good indication of levels of use made of different parks within a given urban area. Again, the numbers of house sales boards erected in an area over a given period of time could prove to be a useful indicator of residential mobility.

121

Mechanical and Electronic Devices

Mechanical and electronic devices may be used simply as means of recording data; or as supplements to other methods of research; or, indeed, as alternative to other methods. As simple recording devices they are most valuable in providing measures of incidence, attendance and the like. The adaptation of devices used for measuring the numbers of cars passing over a given stretch of road, so as to record, say, the number of footsteps passing over a stile, has been employed in a number of studies. The survey of visitors to the Wye and Crundale Nature Reserve utilized such devices.[28] In their present state of development, however, these devices are of relatively limited use – mainly as counters which provide overall totals for the numbers of persons carrying out some particular act. They can tell us very little more about these persons.

Mechanical and electronic devices used as supplements and alternatives to other research methods are, however, of greater value. Hidden cameras and microphones have been used, both as supplements and alternatives to observation, in many studies in the United States. They have been particularly useful in child psychology. Weir made audiotape recordings of a two-and-a-half-year-old child falling asleep, thereby obtaining an insight into language learning processes which it would have been impossible to obtain by simple observation.[72] Potential uses in planning are, for example, the making of films of children at play which would provide a picture over time of the volume and kinds of use made of different pieces of equipment on children's playgrounds. Similarly, films of activities at municipal swimming pools have provided data about the numbers of people actually in the water at any one time as compared with the numbers on surrounding areas.

One of the major problems encountered in the use of mechanical and electronic devices is the difficulty of securing a satisfactory sample of data. If the recording device is located in a single place, such as a camera or counter at a picnic site, then the investigator is forced to accept the data obtained from the persons who pass this spot. If, on the other hand, the recording device is mobile, such as a microphone hidden in a hearing aid, then he exercises greater choice in his search for respondents, and is able, if he wishes, to make an attempt to obtain a meaningful sample of respondents.

Insufficient is yet known about the use of mechanical and electronic devices on a large scale in social research. They are, clearly, very valuable as counters for establishing 'populations' and totals of various kinds. But their use for more analytical research has been confined largely to the field of psychological investigations, where the problems of sampling present few difficulties. They have not been applied systematically to problems in social research. Furthermore, there could be difficulties of a procedural, political or moral kind, in so far as these devices could be interpreted as 'snooping' by the general public. Films and recordings could be considered as akin to telephone tapping – an unwarranted intrusion of privacy.

Documents

Documentary sources are of two main kinds: *continuous records* of the community or sections of the community; and *discontinuous records* of one kind and another. There is a wide variety of documentary sources – for example, birth, marriage, and death records, which can be used for studies of fertility patterns over time; city and town directories, which may be useful for studies of residential mobility; water pressure and electricity consumption records, which have been used to assess level of television viewing in America and Britain; sales records; records of job turnovers; and so on. There are at least two major sources of bias in such documents – *selective deposit* and *selective survival*. The further back in time the investigator goes in his search for documents, the greater these sources of bias are likely to become (cf. problems in archaeological research). The major characteristics of documentary sources are, however, that they have been produced by someone other than the researcher, for someone other than him, and, usually, for some purpose other than his. This is true of both continuous and discontinuous records.

Continuous records offer a large mass of data for many areas of social research. They are relatively cheap to obtain, easy to sample, and population restrictions associated with them are often known and, therefore, capable of control. Their main advantage is that they create opportunities for study over time. Their main disadvantages are, of course, the twin problems of selective deposit and selective survival. Data may vary selectively over time and across different geographical areas. A particular difficulty is to assess the effects of extraneous factors, such as administrative and definitional changes. Records of expenditure upon entertainment for example, may be non-comparable over time; that is, they may be affected by changes in methods of recording or by alterations in the definition of 'entertainment'.

Discontinuous records are of two main kinds: *institutional records* and *personal documents*. Both kinds are potential substitutes for

123

direct observation and surveys. They are often more difficult (and more costly) to acquire than continuous public records, but they usually provide a gain in specificity of content.

Particular emphasis has been given to the use of personal documents in social research during the past twenty years or so. These provide a richness of detail which cannot be obtained from sales records, institutional records, continuous records, direct observation and survey methods. They are, however, of little value for predictive purposes; and it is usually extremely difficult to get a representative sample of them. They are of two main kinds – *existing documents* which the independent investigator is able to utilize; and *elicited documents*, which are produced at the instigation of the investigator by the individuals being studied. Existing documents can be in the form of autobiographies, diaries and letters. Elicited documents are usually in the form of diaries and inventories of various kinds.

The great merit of *autobiographies* is that they give data about that part of an individual's life which is hidden from the objective measures of the social scientist. But, in using an autobiography, it is essential for the investigator to distinguish between the record of actual experience and the interpretations placed upon these experiences by the writer.

The spontaneous, intimate *diary* is the personal document *par excellence*. It is usually written under the immediate influence of experience and, for this reason, is particularly effective in capturing changes of mood. On the other hand, it may neglect calm and happy periods of life as being 'uneventful'. It may also lack continuity and it usually takes much for granted.

The use of *letters* for social research has many of the advantages and disadvantages associated with the use of diaries. There is, however, the additional complication caused by the need to consider the personality of the recipient as well as that of the writer.

Inventories fall between diaries and questionnaires. The main kind that might prove useful in planning studies is the *time-budget diary*. This fulfils a role in the study of time expenditure analogous to that performed by the budget book in the more usual money-expenditure survey. In choosing the time intervals which the subject is to be asked to record, a balance must be struck between the burden to be imposed upon the subject and the amount of detail required. A choice may also have to be made between showing the amount of time spent on various activities over a period of several days, and calculating an 'average day'. To achieve the greatest possible accuracy, explanations and instructions from the investigator should

124

be very clear and the time of recording must be clearly established.*

Allport[3] has identified eleven objections to the use of personal documents for research purposes – three of which, he suggests, are irrelevant, five of which are true only under certain circumstances, and three of which are true in virtually all cases. They are as follows:

I. Personal Documents are too subjective	Irrelevant
II. Their validity cannot be estimated	
III. They are saturated with 'mood'.	
IV. They constitute an unrepresentative sample	True only
V. They are oversimplified by the writer	under certain
VI. They may include deliberate deception	circumstances
VII. They may include unintentional deception	
VIII. The writer is blind to his own motives.	
IX. The writer is fascinated with style	Usually true
X. They may include errors of memory	
XI. They involve implicit selection and, so, bias.	

While this analysis is true in part, it starts from the negative point of attempting to disprove objections to personal documents. The value of personal documents in social research may be cast differently by considering, first, their advantages and, then, their disadvantages.

The advantages of personal documents are, first, that they may often be the only way we have of following through the subject of interest over a long period of time; second, that they usually provide a richness of detail not obtainable by other research methods; third, that they enable the investigation to cover both time and distance; fourth, they are probably at their most useful when unsolicited, thus giving an entirely objective insight into the characters of particular individuals; fifth, they are often useful as the final stage of a multi-phase sample study, when they can often yield data which may have been withheld by respondents during the earlier parts of the study; and sixth, elicited documents in particular may be valuable as adjuncts to interview and postal surveys.

Against this are the disadvantages that, with existing documents, the sample is unrepresentative (but not necessarily with elicited documents); that comparability between individuals is difficult even with elicited questionnaires and inventories; that personal docu-

---

* For a detailed discussion of the merits and limitations of time budget diaries, see T. L. Burton, *Experiments in Recreation Research*, to be published in 1971, especially Chapter 7.

ments, especially time-budget sheets, present major difficulties in coding; that personal documents usually have little value in uncovering motives, attitudes and opinions, which can be elicited by skilful questioning on the part of the interviewer; that they can be misleading because of deliberate or unintended deception – that is, for example, the wish to put oneself in the best possible light; and that they usually have very little predictive value.

Most of the disadvantages of personal documents for use in social research are relative rather than absolute – that is, they are true in certain circumstances only. Individual documents should, therefore, be tested to establish the extent to which these disadvantages exist. A number of tests of credibility, of which there are two main kinds, can be applied to isolate those disadvantages which do exist.

External tests of credibility have to do with procedures for distinguishing between a hoax or misrepresentation and a genuine document. Various kinds of chemical and electronic tests have been devised for assessing the validity of ancient (and modern) documents. Handwriting experts can also be used for this purpose. For planning research purposes, however, external tests of credibility are likely to be of minor significance in relation to internal tests.

Internal tests of credibility aim at establishing the degree to which the *content* of a document is credible. There are four main tests: (I) was the writer able to tell the truth?; (II) was he willing to tell the truth?; (III) is he accurately reported?; and (IV) is there any external corroboration (such as supporting evidence from other documents of one kind or another)? It may often be impossible to obtain a satisfactory answer to all (or any) of these questions, but the asking of them shows an awareness of the particular problems attached to the use of documents.

Observation

Observation can be defined as the purposeful and selective watching and counting of phenomena as they take place. As a systematic method of research, it should satisfy three main conditions: it should be suitable for investigating the problems in which the researcher is interested; it should be appropriate to the populations and samples that he wishes to study; and it should be reliable and objective. Moser[42] suggests that the method has significant weaknesses on all three of these scores. It is usually suitable for only a small fraction of the subjects the researcher wants to study, since it is confined in time and place (in the sense that it can be carried out no more rapidly than the event itself and only in the place where the event happens). It is often not easily combined with sampling procedures. And it usually gives too much scope to the subjective influences of the individual investigator.

Sampling problems are, indeed, a particularly difficult problem in the use of observation as a research method. If the characteristics of a population are to be inferred from those of a sample, then this sample should (usually) be randomly selected. This is virtually impossible when observation alone is the research method. Subjectivity is also a difficult problem. Observation tends to give too much scope to the subjective influence of the individual investigator. Thus, the investigator may have to observe something of which he is himself a part. Observation is necessarily partial and selective; and even though trained and experienced observers may attempt to separate observation from interpretation, it is likely that their own particular biases and feelings may cause them to observe selectively. This we may term 'the myth of objectivity'. We should, perhaps, recognize that the achievement of observation untinged with inference and interpretation is impossible; and so, work towards a system of controlling and measuring the degree to which inference and interpretation are present, rather than try to eliminate them.

Despite these objections, observation can be a useful research tool. It may be valuable, for example, in situations where subjects are unable or unwilling to provide information to the researcher by other methods. Moreover, even though it may allow the subjective influence of the investigator to affect the research findings, it does at least reduce the possibilities of falsifying the data through the conscious or unconscious desires of subjects to present inaccurate pictures of themselves, or to distort information.

Observation can be particularly useful as a complementary technique to interview surveys. It can be applied, for example, to indicate the respondent's style of life, his non-verbal responses in relation to his verbal ones and to check actual behaviour against the respondent's report of his behaviour. The method is particularly valuable for studying small communities and institutions in action and for assessing how people live and how they behave in given situations. It has been used with considerable success in a number of studies of small communities and 'closed' groups within society.[26]

A notable recent instance of one observation-based study has been that of H. J. Gans at Levittown, a New Jersey suburb developed by a single builder and now called by the original name of the township, Willingboro.[19] Gans lived there for the first two years of its existence (1958–60), to find out how a new community comes into being, how people change when they leave the city and how they live and politic in suburbia. His actual research role was more an observer's than a participant's. Becoming a participant in one group effectively excludes a researcher from knowledge of what is happening in competing or opposing groups; Gans decided to participate only in the life of his own block and as a member of the general public at meetings; otherwise his role was that of an observer and informal interviewer. A most interesting study of an American suburb resulted.

A number of different observational methods have been identified by various research workers – participant and non-participant, controlled and uncontrolled, simple and systematic. These descriptions are, however, merely slight variations of similar concepts. Despite the wealth of terms, we can identify only two basic and distinct groups of methods: participant/simple-uncontrolled methods; non-participant/systematic-controlled methods. The former refers to situations which the observer has witnessed or in which he has taken part on an entirely random basis. The investigator's observations are not checked by other observers or against a set of specific items to be noted down; and the observations do not form part of a detailed outline of experiments. The latter refers to situations in which the role of the investigator is not detectable and, therefore, the naturalness of the situation is not disturbed. In practice it is rarely possible to obtain a true nonparticipant observation; it usually means, in fact, quasi-participation.

The participant observer shares in the lives and activities of the persons or groups that he is studying, observing what is happening about him, but supplementing this with conversations, interviews and studies of records. His experiences are unique – so that a second investigator would not be able to record the same data.

The major advantage of participant/simple-uncontrolled observation is that the data are obtained at first hand. Indeed, the outsider may often obtain additional information simply because he is an outsider. Again the technique could be a means of obtaining a balance between purely behaviouristic investigation and more 'meaningful' investigation into motivations and aspirations.

The chief disadvantage of participant/simple-controlled observation is, however, that the investigator may become identified, symbolically or emotionally, with one particular section of the community or group which he is studying. This can lead to his being refused access to other groups. A further problem arises in as much as the critical data that the investigator is seeking are variable over time. The observer participant method is particularly limited whenever data relating to behaviour over time or frequency of behaviour are

128

required. Again, the observer can be variable over time. Another problem arises from population variation: this is often significant, but sampling is extremely difficult if not impossible to carry out. Difficulties also arise over distinctions between what the investigator observes, what he thinks he observes and what he reports that he observes.

While much interesting material has been obtained by the participant observer technique, it is generally true that discrepancies in observations by different observers at the same time and the same observer at different times can seldom be reconciled without the use of more objective techniques. This leads to a consideration of non-participant/systematic-controlled observation. There are at least three requirements for this; careful definition of the units (that is, persons) to be observed and the units to be ignored; selection of the pertinent data about these units which are to be observed and recorded; standardization as far as possible, of the conditions of observation (that is, time, place, persons, and so on).

One significant type of controlled observation has been a situation in which subjects have been told that they were being observed but either they were not told the reasons for observation, or they were given false reasons for it. The objective here is to avoid, if possible, the introduction of bias into the data which could arise by reason of the observer's presence. For example, the presence of an observer in a children's playground could significantly alter the normal pattern of play behaviour, since the children might set out to 'impress' the observer. By drawing the children's attention to the fact that they are being observed but not informing them of the reasons for it, the investigator would randomize the observer-bias. By giving them false reasons for the observation, he could direct potential bias along lines which he assumes (or knows) will not significantly affect those aspects of behaviour in which he is interested.

Non-participant/systematic-controlled observation has several advantages over the simple-uncontrolled variety. If often happens that the behaviour being investigated occurs so infrequently that uncontrolled observation would require an inordinate amount of effort on the part of the investigator, gathering a mass of data only a small part of which will be directly pertinent to the study. Also, it provides the investigator with greater control over the occurrence of events which will stimulate the natural behaviour he wishes to study. Against this, however, there is the disadvantage that, as with simple observation, the investigator can identify with one particular section of the group or community which he is studying. In addition, there is always the difficulty arising from the variability of the observer

I

himself; the difficulty of distinguishing between true observation and inference.

The purpose in outlining these non-survey research techniques has been to show the non-specialist research investigator that it is not always necessary to carry out his research by means of a sample survey. The latter is usually a costly exercise, in both time and money. In many cases this will have to be accepted, since a survey will be the only way of obtaining the data that are required. But sometimes the required information can be obtained by other methods. Particularly useful alternatives will sometimes be Census data, local authority housing records, electoral registers, and so on. If the data that the planner needs are fairly simple and relate to totals of persons within fairly large and homogeneous categories, then institutional records will often be adequate sources. The essential point to be made is that the investigator should always consider first whether the data which he needs are not already in existence – if not in the form in which he needs them, at least in a form which he can adapt. Only when he has thoroughly exhausted all practical possibilities of existing sources should he consider collecting his own data. Even then, he may find that a technique such as observation will be a sufficient, and better, use of resources, than a sample survey.

Bibliography

1 ABRAMS, M., *Education, Social Class and Reading of Newspapers and Magazines*, Institute of Practitioners in Advertising, 1966.
2 ALLEN, D. E., *British Tastes*, Hutchinson, 1968.
3 ALLPORT, G. W., *The Use of Personal Documents in Psychological Science*, Social Science Research Council (USA), 1952.
4 ATKINSON, J., *A Handbook for Interviewers*, Government Social Survey, H.M.S.O., 1968.
5 BURGESS, E. W., 'The Growth of the City', in *The City*, PARK, R. E., *et al.*, (eds), University of Chicago Press, 1925.
6 BURTON, T. L., and NOAD, P. A., *Recreation Research Methods: A Review of Recent Studies*, Centre for Urban and Regional Studies, Occasional Paper No. 3, 1969.
7 CANTRIL, H., *Gauging Public Opinion*, Princeton University Press, 1947.
8 CHAPMAN, D., 'Age, Sex, Marital Status, Birthplace', Papers of the British Sociological Association Working Party on Comparability of Data, Set 1, Paper 1, 1968, Unpublished.
9 CHERRY, G. E., 'Overcrowding in Cities', *Official Architecture and Planning*, 32, 3, March 1969, pp. 287–90.
10 CHERRY, G. E., *Town Planning in its Social Context*, Leonard Hill Books, 1970.
11 CLAUSEN, J. A., and FORD, R. N., 'Controlling Bias in Mail Questionnaires', *Journal of the American Statistical Association*, 42, 1947, pp. 497–511.
12 *Report of the Committee on Higher Education 1961–63* (Robbins Report), Cmnd. 2601, London, H.M.S.O., 1964.
13 COUTER, T., and DOWNHAM, J. S., *The Communication of Ideas*, Chatto and Windus, 1954.
14 COOLEY, C. H., *Social Organisation*, New York, 1909.
15 Council for Children's Welfare, *The Playground Study*, 1967.
16 DUHL, L. J., 'The Parameters of Urban Planning' in ANDERSON, S., (ed.), *Planning for Diversity and Choice*, M.I.T. Press, 1968.
17 FIREY, W., *Land Use in Central Boston*, Harvard University Press, 1947.
18 GALLUP, G., 'The Quintamensional Plan of Question Design', *Public Opinion Quarterly*, Fall 1947, pp. 385 ff.
19 GANS, H. J., *The Levittowners*, Allen Lane the Penguin Press, 1967.
20 GANS, H. J., *People and Plans*, Basic Books Inc., 1969.
21 GEDDES, P., *Cities in Evolution*, 1915.
22 GITTUS, E., 'Income', in STACEY, M., (ed.), *Comparability in Social Research*, Heinemann Educational Books, 1969.
23 Government Social Survey, *Some Data Relating to Refusals and Refusal-Prone Interviewers Based on Performance on Routine Inquiries*, Research Paper M.49, 1952.

I*

24 Government Social Survey, *Analysis and Interpretation*, Methodological Series M.68, 1953.
25 GRAY, P. G., and CORLETT, T., 'Sampling for the Social Survey', *Journal of the Royal Statistical Society*, A, 113, pp. 150–206.
26 GRAY, P. G., and HARRIS, A., *Some Notes on Postal Checks*, Government Social Survey, Research Paper M.44, 1951.
27 HALL, J., and JONES, D. C., 'The Social Grading of Occupations', *British Journal of Sociology*, 1, 1, 1950.
28 HAMMOND, E. C., Survey of Visitors to Wye and Crundale National Nature Reserve, Nature Conservancy, 1965, Unpublished.
29 HAMMOND, E., *An Analysis of Regional Economic and Social Statistics*, Rowntree Research Unit, University of Durham, 1968.
30 HANCOCK, J., 'An Experimental Study of Four Methods of Measuring Unit Costs of Obtaining Attitudes Towards Retail Stores', *Journal of Applied Psychology*, 1940, pp. 213–30.
31 HYMAN, H. H., *et al.*, *Interviewing in Social Research*, University of Chicago Press, 1955.
32 HOYT, H., *The Structure and Growth of Residential Neighborhoods in American Cities*, Federal Housing Administration, 1939.
33 KEMSLEY, W. F. F., *Family Expenditure Survey*, handbook on the sample, fieldwork, and coding procedures, Government Social Survey, H.M.S.O., 1969.
34 KINSEY, A. C., *et al.*, *Sexual Behaviour in the Human Male*, W. B. Saunders Company, 1948.
35 KISH, L., *Survey Sampling*, John Wiley and Sons, 1967.
36 KLUCKHOHN, F. R., 'The Participant-Observer Technique in Small Communities', *American Journal of Sociology*, 46, 1940, pp. 331–43.
37 LEE, T., 'The Optimum Provision and Siting of Social Clubs', *Durham Research Review*, 14, September 1963.
38 Market Research Society, *Social Class Definition in Market Research*, London, 1963.
39 Ministry of Labour, *Family Expenditure Survey: Report for 1966*, H.M.S.O., 1967.
40 Ministry of Labour, *Standard Industrial Classification*, H.M.S.O., 1958.
41 Ministry of Social Security, *Circumstances of Families*, H.M.S.O., 1967.
42 MOSER, C. A., *Survey Methods in Social Investigation*, Heinemann, 1958.
43 MOSER, C. A., and STUART, A., 'An Experimental Study of Quota Sampling', *Journal of the Royal Statistical Society*, A, 116, 1953, pp. 349–405.
44 County Borough of Newcastle upon Tyne, Planning Department, *Evening Leisure Survey and Theatre Survey*, 1967, Unpublished.
45 County Borough of Newcastle upon Tyne, Planning Department, *Parks Survey*, 1967, Unpublished.

46 North Regional Planning Committee, *Mobility and the North*, 1967.

47 OPPENHEIM, A. N., *Questionnaire Design and Attitude Measurement*, Heinemann, 1966.

48 PAHL, R. E., (ed.), *Readings in Urban Sociology*, Pergamon Press, 1968.

49 PARTEN, M. B., *Surveys, Polls and Samples*, Harper and Row, 1950.

50 PAYNE, S. L., *The Art of Asking Questions*, Princeton University Press, 1951.

51 General Register Office, *Census 1966*, H.M.S.O., 1966.

52 General Register Office, *Classification of Occupations*, H.M.S.O., 1966.

53 REUSS, C. F., 'Differences Between Persons Responding and Not Responding to a Mailed Questionnaire', *American Sociological Review*, 8, 1943, pp. 433–38.

54 SAWYER, H. G., *The Meaning of Numbers*, Speech given to the American Association of Advertising Agencies, 1961.

55 SEITZ, R. M., 'How Mail Surveys May be Made to Pay', *Printers' Ink*, 209, 1944, pp. 17–19.

56 SHUTTLEWORTH, F. K., 'A Study of Questionnaire Technique', *Journal of Educational Psychology*, 22, 1931, pp. 652–58.

57 SLETTO, R. F., 'Pretesting of Questionnaires', *American Sociological Review*, 5, 1940, pp. 193–200.

58 SMITH, H. L., and HYMAN, H. H., 'The Biasing Effect of Interviewer Expectations on Survey Results', *Public Opinion Quarterly*, 14, pp. 491–506.

59 STACEY, M., *Tradition and Change: A study of Banbury*, Oxford University Press, 1960.

60 Survey Research Center, *Manual for Coders: Content Analysis*, University of Michigan Institute for Social Research, 1961.

61 SUCHMAN, E. A., and MCCANDLESS, B., 'Who Answers Questionnaires?', *Journal of Applied Psychology*, 24, 1940, pp. 758–69.

62 Town Planning Institute, *Planning Research*, 3rd edition, 1968.

63 United Nations Sub-Commission on Statistical Sampling, *Recommendations Concerning the Preparation of Reports on Sampling Surveys*, New York, 1950.

64 UNWIN, R., *Nothing Gained by Overcrowding*, Garden Cities and Town Planning Association, 1912.

65 University of Birmingham, Centre for Urban and Regional Studies, *Birmingham Recreation Planning Study*, In Progress.

66A University of Keele – British Travel Association, *Pilot National Recreation Survey*, Report No. 1, 1967; Report No. 2, 1969.

66B SILLITOE, K. K., *Planning for Leisure*, Government Social Survey, H.M.S.O., 1969.

67 UTTING, J. E. G., 'An Inquiry into the Economic Circumstances of Old Age', *Journal of the Royal Statistical Society*, A, 119, 1956.

68 WARNER, W. L., *Social Class in America*, Science Research Association Inc., Chicago, 1949.

133

69 WATSON, R., 'Investigations By Mail', *Market Research*, 5, 1937, pp. 11–16.
70 WEBB, E. J., *et al.*, *Unobtrusive Measures: Nonreactive Research in the Social Sciences*, Rand McNally, 1966.
71 WEINBERG, A, 'Education', in STACEY, M., (ed.), *Comparability in Social Research*, Heinemann Educational Books, 1969.
72 WEIR, R. H., *Language in the Crib*, Mouton, 1963.
73 WEBER, M., quoted in COX, O. C., 'Max Weber on Social Stratification: A Critique', *American Sociological Review*, 15, 1950.
74 White Paper, *A Policy for the Arts*, Cmnd. 2601, H.M.S.O., 1965.
75 WILKINS, L. T., *Prediction of the Demand for Campaign Stars and Medals*, The Social Survey, 1949.
76 WILLMOTT, P., and YOUNG, M., 'Social Grading by Manual Workers', *British Journal of Sociology*, 7, 4, 1956.
77 YATES, F., *Sampling Methods for Censuses and Surveys*, Griffin, 3rd edition, 1960.

INDEX

Milton Keynes UK
Ingram Content Group UK Ltd.
UKHW031152141024
449569UK00024B/864